Lecture Notes in Bioinformatics

T0238270

Edited by S. Istrail, P. Pevzner, and M. Waterman

Editorial Board: A. Apostolico S. Brunak M. Gelfand
T. Lengauer S. Miyano G. Myers M.-F. Sagot D. Sankoff
R. Shamir T. Speed M. Vingron W. Wong

Subseries of Lecture Notes in Computer Science

Springer
Berlin
Heidelberg
New York
Hong Kong
London
Milan
Paris
Tokyo

Concettina Guerra   Sorin Istrail (Eds.)

Mathematical Methods for Protein Structure Analysis and Design

C.I.M.E. Summer School
Martina Franca, Italy, July 9-15, 2000
Advanced Lectures

 Springer

Series Editors

Sorin Istrail, Celera Genomics, Applied Biosystems, Rockville, MD, USA
Pavel Pevzner, University of California, San Diego, CA, USA
Michael Waterman, University of Southern California, Los Angeles, CA, USA

Volume Editors

Concettina Guerra
Università degli Studi di Padova
Dipartimento di Ingegneria dell'Informazione
via Gradenigo 6a, 35131 Padova, Italy
E-mail: guerra@dei.unipd.it

Sorin Istrail
Celera Genomics, Applied Biosystems
45 West Gude Drive, Rockville, MD 20850, USA
E-mail: sorin.istrail@celera.com

Cataloging-in-Publication Data applied for

A catalog record for this book is available from the Library of Congress.

Bibliographic information published by Die Deutsche Bibliothek.
Die Deutsche Bibliothek lists this publication in the Deutsche Nationalbibliografie;
detailed bibliographic data is available in the Internet at <http://dnb.ddb.de>.

CR Subject Classification (1998): J.3, F.2, H.2, G.2, I.3.5, I.4

ISSN 0302-9743
ISBN 3-540-40104-0 Springer-Verlag Berlin Heidelberg New York

Springer-Verlag Berlin Heidelberg New York
a member of BertelsmannSpringer Science+Business Media GmbH

http://www.springer.de

© Springer-Verlag Berlin Heidelberg 2003
Printed in Germany

Typesetting: Camera-ready by author, data conversion by Boller Mediendesign
Printed on acid-free paper   SPIN: 10927403   06/3142   5 4 3 2 1 0

Preface

The papers collected in this volume reproduce contributions by leading scholars to an international school and workshop which was organized and held with the goal of taking a snapshot of a discipline under tumultuous growth. Indeed, the area of protein folding, docking and alignment is developing in response to needs for a mix of heterogeneous expertise spanning biology, chemistry, mathematics, computer science, and statistics, among others.

Some of the problems encountered in this area are not only important for the scientific challenges they pose, but also for the opportunities they disclose in terms of medical and industrial exploitation. A typical example is offered by protein-drug interaction (docking), a problem posing daunting computational problems at the crossroads of geometry, physics and chemistry, and, at the same time, a problem with unimaginable implications for the pharmacopoeia of the future.

The school focused on problems posed by the study of the mechanisms behind protein folding, and explored different ways of attacking these problems under objective evaluations of the methods. Together with a relatively small core of consolidated knowledge and tools, important reflections were brought to this effort by studies in a multitude of directions and approaches. It is obviously impossible to predict which, if any, among these techniques will prove completely successful, but it is precisely the implicit dialectic among them that best conveys the current flavor of the field. Such unique diversity and richness inspired the format of the meeting, and also explains the slight departure of the present volume from the typical format in this series: the exposition of the current sediment is complemented here by a selection of qualified specialized contributions.

The topics covered in this volume pinpoint major issues arising in the development and analysis of models, algorithms and software tools centered on the structure of proteins, all of which play crucial roles in structural genomics and proteomics. The study of 3D conformations and relationships among proteins is motivated by the belief that the spatial structure, more than the primary sequence, dictates the function of a protein. The largest repository of

3D protein structures is the Protein Data Bank (PDB), currently containing about 17,000 proteins. The PDB has experienced a sustained growth and is expected to continue to grow at an increasing pace in the near future. The available structures are classified into a relatively small number of families and folds, according to their three-dimensional conformation. While the number of proteins will continue to grow, it is widely believed that the number of new folds will remain relatively stable. Structural comparisons involving these structures are at the core of docking and the classification of proteins and sub-aggregates, and motif searches in sequence and protein databases, and ultimately they contribute to understanding the mechanics of folding in living organisms.

The first three chapters of this volume contain material that was presented at the school. The chapter entitled "Protein Structure Comparison: Algorithms and Applications," by G. Lancia and S. Istrail, focuses on the algorithmic aspects and applications of the problem of structure comparison. Structure similarity scoring schemes used in pairwise structure comparison are discussed with respect to the ability to capture the biological relevance of the chemical and physical constraints involved in molecular recognition. Particular attention is paid to the measures based on *contact map* similarity.

The chapter "Spatial Pattern Detection in Structural Bioinformatics," by H.J. Wolfson, discusses the task of protein structural comparison as well as the prediction of protein-protein, protein-DNA or protein-drug interaction (docking). Different protein shape representations are used in biological pattern discovery. The paper discusses the shape representations best suited to each computational task, then outlines some rigid and flexible protein structural alignment algorithms, and discusses the issues of rigid bound versus unbound and flexible docking.

The chapter "Geometric Methods for Protein Structure Comparison," by C. Ferrari and C. Guerra, discusses, from a theoretical point of view, geometric solutions to the problem of finding correspondences between sets of geometric features, such as points or segments. After reviewing existing methods for the estimation of rigid transformations under different metrics, the paper focuses on the use of the secondary structures of proteins for fast retrieval of similarity. It also deals with the integration of strategies using different levels of protein representations, from atomic to secondary structure level.

The chapter "Identifying Flat Regions and Slabs in Protein Structures," by M.E. Bock and C. Guerra, presents geometric approaches to the extraction of planar surfaces, which is motivated by the problem of identifying packing regions in proteins.

The two contributions, "OPTIMA: a New Score Function for the Detection of Remote Homologs," by M. Kann and R.A. Goldstein, and "A Comparison of Methods for Assessing the Structural Similarity of Proteins," by D.C. Adams and G.J.P. Naylor, deal with the problem of protein comparison, focusing on different similarity functions for sequence and structure comparison.

The next three papers, "Prediction of Protein Secondary Structure at High Accuracy Using a Combination of Many Neural Networks," by C. Lundegaard, T.N. Petersen, M. Nielsen, H. Bohr, J. Bohr, S. Brunak, G. Gippert and O. Lund, "Self-consistent Knowledge-Based Approach to Protein Design," by A. Rossi, C. Micheletti, F. Seno, A. Maritan, and "Learning Effective Amino-Acid Interactions," by F. Seno, C. Micheletti, A. Maritan and J.R. Banavar, discuss techniques and criteria for protein folding and design.

The paper "Protein structure from solid-state NMR," by J.R. Quine and T.A. Cross, presents a mathematical analysis for solid-state nuclear magnetic resonance (NMR). Finally, the contribution "Protein-like Properties of Simple Models," by Y.-H. Sanejouand and G. Trinquier, focuses on properties relevant to the sequence-structure relationships.

The school was attended by 56 participants from 10 countries. Lectures were given by Prof. Ken Dill, University of California (USA), Prof. Arthur Lesk, University of Cambridge Clinical School (UK), Prof. Michael Levitt, Stanford University School of Medicine (USA), Prof. John Moult, University of Maryland (USA), and Prof. Haim Wolfson, Tel Aviv University (Israel) Invited talks at the workshop were given by Prof. Mary Ellen Bock, Purdue University (USA) and Dr. Andrea Califano, IBM Yorktown (USA).

Concettina Guerra
Sorin Istrail

Contents

Protein Structure Comparison: Algorithms and Applications
Giuseppe Lancia, Sorin Istrail 1
1 Introduction 1
2 Preliminaries 4
3 Applications of Structure Comparisons 6
4 Software and Algorithms for Structure Comparison 11
5 Problems Based on Contact Map Representations 20
6 Acknowledgements 30
References 30

Spatial Pattern Detection in Structural Bioinformatics
Haim J. Wolfson.. 35
1 Introduction 35
2 Protein Shape Representation 37
3 Protein Structural Alignment 38
4 Protein-Protein Docking...................................... 46
5 Summary.. 52
References 53

Geometric Methods for Protein Structure Comparison
Carlo Ferrari, Concettina Guerra 57
1 Introduction 57
2 Protein Description 59
3 Structural Comparison: Problem Formulation 62
4 Representation of Rigid Transformations 63
5 Determination of 3D Rigid Transformations..................... 68
6 Geometric Pattern Matching.................................. 71
7 Indexing Techniques 73
8 Graph-Theoretic Approaches.................................. 76
9 Integration of Methods for Protein Comparison Using Different
   Representations 77

10 Conclusions.. 78
11 Acknowledgements 79
References 79

Identifying Flat Regions and Slabs in Protein Structures
Mary Ellen Bock, Concettina Guerra 83
1 Introduction 83
2 A Geometric Algorithm 85
3 An Improved Geometric Algorithm 87
4 Hough Transform.. 88
5 Performances of the Two Algorithms........................ 90
6 Plane Detection in Proteins................................. 92
7 Acknowledgements 95
References 96

OPTIMA: A New Score Function for the Detection of Remote Homologs
Maricel Kann, Richard A. Goldstein.............................. 99
1 Abstract 99
2 Introduction 99
3 Methods 100
References 107

A Comparison of Methods for Assessing the Structural Similarity of Proteins
Dean C. Adams, Gavin J.P. Naylor 109
1 Introduction 109
2 The DALI Algorithm 109
3 The Root Mean Square Algorithm 110
4 Geometric Morphometrics 111
5 Comparison of Methods 111
6 Discussion 113
References 114

Prediction of Protein Secondary Structure at High Accuracy Using a Combination of Many Neural Networks
Claus Lundegaard, Thomas Nordahl Petersen, Morten Nielsen, Henrik Bohr, Jacob Bohr, Søren Brunak, Garry Gippert, Ole Lund.....117
1 Summary... 117
2 Introduction 117
3 Methods 118
4 Results.. 119
References 121

Self-consistent Knowledge-Based Approach to Protein Design
Andrea Rossi, Cristian Micheletti, Flavio Seno, Amos Maritan 123
1   Introduction 123
2   The Design Strategy 124
3   Results and Discussion 125
4   Summary 127
References 128

Protein Structure from Solid-State NMR
John R. Quine, Timothy A. Cross 131
1   Discrete Curves 131
2   Tensors and NMR 133
3   Structure from Orientational Constraints 134
4   Acknowledgment 136
References 136

Learning Effective Amino-Acid Interactions
Flavio Seno, Cristian Micheletti, Amos Maritan, Jayanth R. Banavar . . 139
1   Introduction 139
2   Models and Techniques 141
3   Results 142
4   Conclusions 144
References 144

Proteinlike Properties of Simple Models
Yves-Henri Sanejouand, Georges Trinquier 147
1   The 3x3x3 Cubic Lattice Model 147
2   N-Soft-Spheres Models 151
References 152

List of Participants 155

Protein Structure Comparison: Algorithms and Applications

Giuseppe Lancia[1] and Sorin Istrail[2]

[1] Dipartimento Elettronica e Informatica, Via Gradenigo 6/A, Padova, Italia
lancia@dei.unipd.it
[2] Celera Genomics, 45 West Gude Dr., Rockville, MD
sorin.istrail@celera.com

1 Introduction

A protein is a complex molecule for which a simple linear structure, given by the sequence of its aminoacids, determines a unique, often very beautiful, three dimensional shape. Such shape (3D structure) is perhaps the most important of all protein's features, since it determines completely how the protein functions and interacts with other molecules. Most biological mechanisms at the protein level are based on shape-complementarity, so that proteins present particular concavities and convexities that allow them to bind to each other and form complex structures, such as skin, hair and tendon. For this reason, for instance, the drug design problem consists primarily in the discovery of ad hoc peptides whose 3D shape allows them to "dock" onto some specific proteins and enzymes, to inhibit or enhance their function.

The analysis and development of models, algorithms and software tools based on 3D structures are heceforth very important fields of modern Structural Genomics and Proteomics. In the past few years, many tools have emerged which allow, among other things, the comparison and clustering of 3D structures. In this survey we cannot possibly recollect all of them, and even if we tried, our work would soon be incomplete, as the field is dynamically changing, with new tools and databases being introduced constantly. At the heart of all of them, there is the Brookhaven Protein Data Bank, the largest world's repository of 3D protein structures.

Two typical problems related with protein structure are *Fold Prediction* and *Fold Comparison*. The former problem consists in trying to determine the 3D structure of a protein from its aminoacid sequence alone. It is a problem of paramount importance for its many implications to, e.g., medicine, genetic engineering, protein classification and evolutionary studies. We will not discuss

C. Guerra, S. Istrail (Eds.): Protein Structure Analysis and Design, LNBI 2666, pp. 1-33, 2003.

the Fold Prediction problem here, if not marginally, but we will instead focus on algorithms and applications for protein fold (structure) comparison. The following are some of the reasons why the Fold Comparison problem is also extremely important:

- *For determining function.* The function of a new protein can be determined by comparing its structure to some known ones. That is, given a set of proteins whose fold has already been determined and whose fuction is known, if a new one has a fold highly similar to a known one, then its function will similar as well. This type of problems imply the design of search algorithm for 3D databases, where a match must be based on structure similarity. Analogous problems have already been studied in Computational Geometry and Computer Vision, where a geometric form or object has to be identified by comparing it to a set of known ones.
- *For clustering.* Given a set of proteins and their structures, we may want to cluster them in families based on structure similarity. Furthermore, we may want to identify a consensus structure for each family. In this case, we would have to solve a multiple structure alignment problem.
- *For assessment of fold Predictions.* The Model Assessment Problem is the following: Given a set of "tentative" folds for a protein, and a "correct" one (determined experimentally), which of the guesses is the closest to the true? This is, e.g., the problem faced by the CASP (Critical Assessment of Structure Prediction) jurors, in a biannual competition where many research groups try to predict protein structure from sequence. The large number of predictions submitted (more than 10,000) makes the design of sound algorithms for structure comparison a compelling need. In particular, such algorithms are at the base of CAFASP, a recent Fully Automated CASP variant.

The problem of Structure Similarity/Alignment determination is in a way analogous to the Sequence Alignment problem, but the analogy is only superficial and it breaks down when it comes to their complexity. There is a dramatic difference between the complexity of sequence alignment and structure alignment. As opposed to the protein sequence alignment, where we are certain that there is a unique alignment to a common ancestor sequence, in structure comparison the notion of a common ancestor does not exist. Similarity in folding structure is due to a different balance in folding forces, and there is not necessarily a one-to-one correspondence between positions in both proteins. In fact, for two homologous proteins that are distantly related, it is possible for the structural alignment to be entirely different from the correct evolutionary alignment [14].

By their nature, three-dimensional computational problems are inherently more complex than the similar one-dimensional ones for which we have more effective solutions. The mathematics that can provide rigorous support in understanding models for structure prediction and analysis is almost nonexistent, as the problems are a blend of continuous, geometric- and combinatorial, discrete-mathematics. Not surprisingly, various simplified versions of the structure comparison problems were shown NP-complete [17, 18].

The similarity of two structures can be accessed on whole proteins or on local domains (e.g., in a clearly defined multi–domain target). This is analogous to global vs local sequence alignments, and different techniques and similarity scores should be used for the two cases.

In this article we focus on the algorithmical aspects and applications of the problem of structure comparison. We will pay particular attention to the measures based on *contact map* similarity. For a more general treatment of different algorithms and measures, the reader is referred to a good survey by Lemmen and Lengauer [30].

An inherent difficulty in pairwise structure comparison is that it requires a structure similarity scoring scheme that captures biological relevance of the chemical and physical steric constraints involved in molecular recognition. In this respect, contact maps seem to have many advantages over other representations. The information of contacts is relatively simple but complete, and it provides a reliable first–approximation to the overwelming complexity of the problem.

Among the classical measures (non contact map–based) for structure comparison, there is the RMSD (Root Mean Square Deviation) which tries to determine an optimal rigid superpositioning of a set of residues of one structure into a predetermined set of corresponding residues of another. The RMSD measure has many recognized flaws. Most notably, it is a global measure which can be a very poor indicator of the quality of a model when only parts of the model are well predicted. In fact, the wrongly predicted regions produce such a large RMSD that it is impossible to see if the model contains "well predicted" parts at all. Despite its drawbacks, the RMSD measure is widely used, and many other measures are based on it. We will see, e.g., the algorithms Max-Sub and GDT, used in CAFASP, the algorithms MNYFIT and STAMP, used by the database HOMSTRAD, 3dSEARCH, used in the database SCOP, the server DALI, and others.

With some notable exception, (e.g. 3dSEARCH) of algorithms whose complexity and properties were analized rigorously and in detail, these algorithms are usually heuristic procedures using some sort of simple local search to optimize a certain objective function. These heuristics are typically based on reasonable assumptions on the properties that the optimal solution should have in order to heavily prune the search space. From a theoretical computer scientist's perspective, the analysis and foundation of these algorithms, if present at all, are often unsatisfactoy, and we remain in doubt that different, more sophisticated approaches (e.g. Branch–and–Bound, or Metaheuristics ad hoc) may be able to compute solutions with a better value of the particular objective function.

Every optimization model set up for a problem in molecular biology is only an arbitrary *representation* of the natural process, and hence it is arguable that its optimal solution is in fact the "right" one. This argument is usually employed to justify the use of simple heuristics instead of any rigorous techniques or provable property of the model. However, after a heuristic optimization, the so–called "global optimum" found is then used to draw conclusions of biological meaning or support some theory on the biological process. Hence, it is important for the quality of the solution to be as good as possible, which justifies the use of sophisticated optimization techniques. It is auspicable therefore an ever increasing cooperation of the community of theoretical computer scientists, applied discrete mathematicians and molecular biologists to address these problems starting from their computational complexity, devising exact and approximation algorithms, studying lower and upper bounds, and designing effective heuristics of many sorts.

The remainder of the paper is organized as follows. We start with a short account on 3D structures, their properties and how they are determined. The rest of the article is divided roughly in two parts. The first part focuses on applications of structure alignment problems, while the second part is centered on algorithms. This second part is in turn divided in two, first talking about measures which are not based on contact maps, and then focusing on contact map problems. Among the applications of structure comparison, we describe some databases based on structure similarity (PDB, SCOP, HOMSTRAD, SLoop, CAMPASS, FSSP) and the problem of protein folding prediction assessment, relevant, e.g., in the CASP and CAFASP competitions. We then describe some of the main algorithms for structure comparison, and in particular those used by some of the applications previously mentioned. We survey RMSD and its variants, DALI, MaxSub, Lgscore, GDT, Geometric Hashing (3dSEARCH), MNYFIT and STAMP. In the final part of the article, we talk about novel algorithms and problems based on the contact map representation of structures. We review the literature and some of our work on exact and heuristic algorithms for the maximum contact map overlap problem. The chapter is closed by a section on possible generalizations and directions for future work.

2 Preliminaries

A *protein* consists of a chain of *aminoacids*, of length varying from about 50 to 3,000 and more. The chemical structure of a single aminoacid is comprised of a carbon atom (called C_α) connected to a carboxyl group and an amine group, a hydrogen atom and a part depending on the specific aminoacid, called the *residue*. The amine group of one aminoacid is linked to the carboxyl group of the next one, giving rise to the linear chain. The *backbone* of the protein is the sequence of C_α atoms, to which the rest of the atoms are attached.

The chemical properties and forces between the aminoacids are such that, whenever the protein is left in its natural environment, it folds to a specific 3-dimensional structure, called its *native*, which minimizes the total free energy. Several experiments have confirmed that two proteins with the same aminoacid sequence have the same 3D structure under natural conditions, and that, if a folded protein is artificially stretched to a chain and then released, it will fold again to the same native.

The 3D structure of a protein is fully specified by the set of (x, y, z)–coordinates of its atoms, with respect to an arbitrary origin. An alternative, weaker description is the set of distances between either all atoms or only the C_αs. The main procedures used to to determine the 3D structure of a protein are *X-ray crystallography* and *Nuclear Magnetic Resonance.* In X–ray crystallography the 3D structure of a molecule is determined by X–ray diffraction from crystals. This technique requires the molecule to be first *crystalized,* at perfect quality; then, its *diffraction pattern,* produced by X–irradiation is studied. This involves the analysis of thousands of spots, each with a position and an intensity. Obtaining good crystals and studying the phases of the waves forming each spot are very complex problems. For these reasons, X–ray crystallography is a very long and delicate process, which may take several months to complete. The result is the set of spacial coordinates of each atom in the structure. For a good description of this technique, see [47] or [5].

Nuclear Magnetic Resonance (NMR) is performed in an aqueous solution, where the molecules tumble and vibrate from termal motion. NMR detects the chemical shift of atomic nuclei with nonnull spin. In order to get an adequate resolution, the molecule must tumble rapidly, which typically limits the size of the proteins to which this technique is applicable. The result of NMR is a set of (estimates of) pairwise distances between the atoms and hence it yields a collection of structures (namely all those compatible with the observed atom–to–atom distances) rather than a single one. According to some studies, the results of NMR are "not as detailed and accurate as that obtained crystallographically" [8].

For a protein we distinguish several levels of structure. At a first level, we have its *primary structure,* given by the monodimensional chain of aminoacids. Subsequent structures depend on the protein fold. The *secondary structure* describes the protein as a chain of structural elements, the most important of which are α–*helices* and β–*sheets.* The *tertiary structure* is the full description of the actual 3D fold, while the *quaternary structure* describes the interaction of several copies of the same folded protein. Correspondingly, when comparing two proteins, one may use algorithms that highlight differences and similarities at each of these levels. Algorithms for the comparison of primary structures are known as *sequence alignment* methods and are beyond the scope of this paper. Here, we will focus mainly on algorithms based on the secondary and tertiary structures.

The input to a generic structure alignment algorithm contains the description of the n–ary structures of two proteins (e.g., for $n = 3$, the set of 3D atomic coordinates of two different molecules). Loosely speaking, the goal of the algorithm is to find a geometrical transformation (typically rigid, made of rotation and translation) which maps a "large" number of elements of the first molecule to corresponding elements of the second. Such algorithms can be of two types, i.e. *sequence order dependent* or *sequence order independent*. In the sequence dependent situation, the mapping must take into account the "identity" of the atoms, i.e. an atom of one structure can only be mapped to the same atom in the other. This way the problem is reduced to a 3D curve matching problem, which is essentially a monodimensional case, and so computationally easier. This approach can be useful for finding motifs preserving the sequence order. A sequence order independent algorithm does not exploit the order of atoms, that is, treats each atom as an anonymous "bead" indistinguishable from the others. This is a truly 3D task, which can detect non–sequential motifs and binding sites. It can be used to search structural databases with only partial information and it is robust to insertion and deletions. Consequently, it is also more challenging computationally.

In the remainder of this survey we will discuss algorithms of both types. The most important sequence order dependent similarity measure is certainly the *Root Mean Square Deviation* (RMSD). This is the problem of rigid superposition of n points (a_i) and n points (b_i), by means of a rotation R and a translation t which minimize $\sum_i |Ra_i + t - b_i|^2$, and can be solved in $O(n)$ time [55] by Linear Algebra techniques. As already mentioned, this measure is not flawless: in fact, the RMSD for a pair of structures almost identical, except for a single, small region of dissimilarity, can be very high.

3 Applications of Structure Comparisons

3.1 Databases Organized on Structure

The *Protein Data Bank* (PDB, [3]) is the most comprehensive single worldwide repository for the processing and distribution of 3D biological macromolecular structure data. As soon as a new 3D structure has been determined, it can be deposited in the PDB, this way making it readily available to the scientific comunity. Currently, the 3D structures of proteins in the PDB are determined mainly by X-ray crystallography (over 80%) and Nuclear Magnetic Resonance (about 16%). The remaining structures are determined by other methods, such as theoretical modeling, which are not experimental in nature, but rather based on computation and/or similarity to known proteins.

In this archive, one can find the primary and secondary structure information, atomic coordinates, crystallographic and NMR experimental data, all annotated with the relevant bibliographic citations, in a suitable text file format. Several viewers are available which interpret this format to give a 3D graphic rendering of the protein.

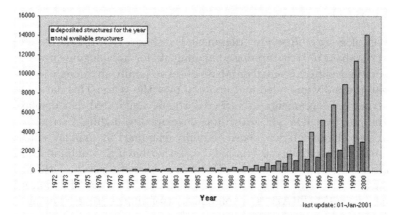

Fig. 1. The growth of the PDB.

The amount of information in the PDB has increased rapidly over the past years, passing from the roughly one thousand structures of 1990 to over 14,000 of today (see Figure 3.1[3]). In order for this information to be usable and accessible, it must be organized into databases which can be queried, e.g., for global/local 3D similarity and/or complementarity, or for finding evolutionary related homologs. A number of such databases has been created in the last decade, each with its own special features. Common to all of them is the use of structure comparison algorithms as the basic tool for clustering and information retrieval. In the remainder of the section we survey a few of the most known, i.e. SCOP, FSSP, HOMSTRAD, SLoop and CAMPASS.

SCOP

In the SCOP (Structural Classification Of Proteins, [42]) database, all proteins of known structure are related to each other according to their structural similarity and evolutionary relationships. The database was created by both manual inspection and the use of algorithms and automated methods. It contains detailed information on all known protein folds, and on all structurally close relatives to a given protein. The SCOP database is also a starting point for the construction of more specific databases, of which we will mention a few later on in the survey.

As of 2000, the database contained information on some 12,000 entries from the PDB, with relative folds, families and superfamilies statistics. A hidden Markov Model for SCOP superfamilies has recently been proposed by Gough *et al* [19]. One of the programs used in the partially automatic classification process in SCOP is 3dSEARCH [51], a procedure based on Geometric Hashing, which is a technique originally developed in the field of Computer Vision. We will shortly describe the technique and the algorithm later in this paper.

---

[3] Figure from `http://www.rcsb.org/pdb/holdings.html`

HOMSTRAD

The HOMologous STRucture Alignment Database (HOMSTRAD, [39]) is a curated database of structure-based alignments for homologous protein families. It originated from a small database of seven family alignments [44], which has been gradually extended and updated over the years. This database provides structural alignments in various formats, annotated and displayed by using the program JOY [38] which was developed to highlight structural features. These alignments have been carefully examined by manual editing and are organized in homologous families, i.e., either having a common evolutionary origin, or a high percentage of sequence identity (most of the families have on average more than 30 percent identity). The classification scheme adopted by HOMSTRAD is a combination of many others, among which those adopted by SCOP. The sequence alignment programs used to help in the classification are BLAST and FUGUE. In HOMSTRAD, the known protein structures are clustered in evolutionary related families, together with a sequence representative of each familiy. These representatives are computed on the basis of common 3D features by using structure alignment programs such as MNYFIT and STAMP, described later in this article. The database also provides superimposed structures and links to other databases and alignment/comparison programs. The use of HOMSTRAD and JOY is suitable for comparative modelling and accurate structure alignments, while SCOP is more suited for the hierarchical classification of protein structures.

SLoop

The basis of the SLoop database [7] is in the work of Donate et al. [11], who described a classification of protein loops and loop regions. SLoop is a database of super secondary fragments, in which proteins are clustered according to the similarity of their secondary structures, in particular the lenght and type of loop regions. The clustering of structures derives form a two–stage process. First, the program SSTRUC is used to identify the loop regions, and each loop is separated into groups according to the length and type of the bounding secondary structures. Then, the loops within each group are pairwise superimposed, the relative similarities are computed and then used in a classical clustering scheme.

CAMPASS

The CAMbridge database of Protein Alignments organised as Structural Superfamilies (CAMPASS, [53]) database is a collection of structure–based sequence alignments of protein(domain)s belonging to a set of superfamilies. Currently, it is restricted to proteins sharing the same topology, and an arbitrary cut-off of 25% sequence identity has been used to eliminate homologous entries within a superfamily. The superfamilies have been chosen according to the SCOP database and the literature. Individual domains within multi-domains proteins have been considered separately. Altough this database is founded on primary structure similarity, it fits within the scope of this survey since for the distantly related members of each superfamily, simple multiple sequence alignment procedures are not appropriate. Hence, the alignment of superfamily members is based on the conservation of structural features like the presence of secondary structures, hydrogen bonding and solvent accessibility. For this type of alignments, the program COMPARER [49] has been used, which takes into account structural information. In the present version, the database consists of 52 superfamilies for which structure-based sequence alignments are available.

FSSP

The Fold classification based on Structure–Structure alignment of Proteins (FSSP, [26]) database provides all–against–all structure comparison of all current PDB structures with more than 30 residues. The alignment and clusters are updated continuously via the DALI search engine (described later in this article). The classification is fully automatic. As of this writing, the database contains some 2,700 sequence families, out of a population of roughly 25,000 protein structures.

The clustering of all proteins in the PDB is done by first identifying a set of "representative" folds (which are highly dissimilar) and partitioning all the remaining proteins into homologs of the structures in the representative set (where a sequence is considered a homolog of another if they have at least 25% sequence identity). Then, an all–againts–all comparison is performed on the sequences in the representative set, this way inducing a clustering of all structures and, implicitly, their homologs. The FSSP entries include the resulting alignments as well as structure alignments of the representatives and their homologs.

3.2 Fold Prediction and Assessment

The problem of protein fold prediction is perceived as one of the core questions, albeit extremely complex, for the molecular biology and gene therapy of the 21st century. The amount of time and work required to determine a protein structure experimentally makes the design of faster yet highly reliable methods a compelling need. One of the possibilities which is extensively studied is the creation of algorithms capable of computing the final fold of a protein from its aminoacid sequence. This is a challenging computational problem and the prize for its solution includes, beyond the obvious scientific and medical implications, also relevant economical interests. For this reasons, some companies among which IBM, are devoting a good deal of research into this problem. In 1999 IBM announced a $100 million research project, named *Blue gene* [2], for the development of a system capable of more than one petaflop (10^{15} operations per second) which will tackle the problem of protein folding by simulating the actual physical phenomenon in a massive, parallel, computation.

The computational problem of protein fold prediction is beyond the scope of this paper. For a good introduction to the topic see, e.g., [34]. What is relevant for our purposes is the assessment of the prediction quality, i.e., how can we judge if a prediction is "close enough" to a model? Clearly, we need tools of structure comparison which, given a model determined experimentally, and a prediction of the former, are capable to come up with a *score* (possibly, a real number in the range $[0, 1]$, with 0 meaning completely wrong, 1 meaning completely right).

We now describe some of the situations where this problem occurs and how it is currently solved.

CASP and CAFASP

The Critical Assessment of techniques for protein Structure Prediction (CASP [40, 41]) is an international bi–annual exercise in which various research groups try to predict the 3D structure of some proteins from their aminoacid sequences. The experiment consists of three phases: First the prediction targets are collected from the experimental community; then the predictions are collected from the modeling comunity; finally, the assessment and discussion of the results takes place. Depending on the method used to predict, there are different categories, i.e. Comparative Modeling, Fold Recognition or Threading, Docking and *ab initio* Folding. The rules of the exercise do not forbid human intervention in making the predictions. In fact, the most successfull teams so far are those who can annoverate some of the best experts in the field. Similarly, the assessment of the results is not done automatically, but, again, human expertise is required. Thus, for each prediction category, a group of evaluators is asked to compile a ranking of the teams. Some automatic methods are employed as well, with the caveat that measures such as RMSD and other rigid–body superposition methods have been found unsuited for the recognition of well predicted substructures or domains. Since it has been so far unclear even if a single, objective, quantitative assessment based on just one measure can exist, in the CASP experiments a large number of scores have been employed, but the final word is left to the expert evaluation and human visualization. Some of the assessment techniques are described in [35]. The large number of models submitted (11136 in CASP4) underlines the ever increasing importance of automatic measures.

CAFASP [12] is a Fully Automated CASP, in which human intervention is forbidden. In this competition, held in parallel with CASP, both predictions and evaluations must be performed by unsupervised programs. Although it is widely believed that human-expert predictions and assessments can be significantly more accurate than their automatic counterparts, one of the goals of CAFASP is to push the research for reducing the gap between the two. The main algorithm used in CAFASP for scoring the predictions is MaxSub, described later in this paper. This algorithm was partially validated by scoring some of the CASP3 Fold-Recognition (FR) models and comparing the rankings obtained to the human–expert assessment carried out by CASP jurors. Even if some differences were observed, the top six predicting groups ranked by MaxSub were the same as those ranked by CASP3. On the other hand, this algorithm had very little success in comparing hard FR targets to their models, because it fails to detect local structural similarities. In the words of CAFASP organizers [13], *even for a target with a known fold, the fact that MaxSub scored a prediction with zero does not necessarily imply that the prediction is totally wrong. (...) Because of the relatively low sensitivity of our automated evaluation, a "human expert" is required to learn more about the prediction capabilities of the servers.*

Live Bench

The Live Bench Project [6] is an effort aimed at the continuous benchmarking of protein structure prediction servers. This project implements an automatic large–scale assessment of the performance of publicly available fold–recognition/prediction servers, such as PDB–Blast, GenTHREADER, FFAS, T98-lib, 3D-PSSM and INBGU. The goal is to provide a consistent, automatized framework in which such servers can be evaluated and compared. Every week, all new PDB protein structures are submitted to all participating fold recognition servers. For instance, in the period October 1999–April 2000, 125 targets were used for comparisons, divided into 30 "easy" structures and 95 "hard" ones. The results of the predictions are compared using similar algorithms as in CAFASP, i.e., MaxSub and Lgscore. Note that, differently from CASP and CAFASP, these "predictions" are in fact done after the structures are made known and available on the PDB, and hence the reliability of LiveBench is based on the assumption that the evaluated servers do not use any hidden feature directing the prediction towards the correct answer. One of the main results of the LiveBench project was to recognize that the servers largely disagree in many cases (they were able to produce structurally similar models for only one half of the targets, and for only one third of the targets such models were significantly accurate) but that a "combined consensus" prediction, in which the results of all servers are considered, would increase the percentage of correct assignments by about 50%.

4 Software and Algorithms for Structure Comparison

In this section we review some of the existing algorithms for protein structure comparison. We start with the RMSD measure, which is at the basis of other algorithms described, such as MaxSub, GDT, 3dSEARCH and MNY-FIT. The section includes both sequence order dependent and sequence order independent algorithms. In the latter we find DALI, Lgscore, 3dSEARCH and STAMP.

A general comment which applies to almost all of the algorithms described is a lack of rigorousness from a Theoretical Computer Science point of view. That is, these algorithms are usually heuristic procedures using some sort of local search to optimize a certain objective function. The algorithms that we surveyed are not "approximate" as intended in Combinatorial Optimization (i.e., there is no guarantee that the solution found will be within a certain factor of the optimum, or of a bound to the optimum, which is never studied), but are mostly fairly simple heuristics which use "common sense" to prune the search space. That is, they exploit some reasonable assumptions on the properties that the optimal solution should have (which, however, are not proved to always hold) to discard partial solutions from further considerations. This way of proceeding makes certainly sense for hard problems such as these, and in fact, the ideas on which Branch-and-Bound is based are somewhat similar. However, Branch-and-Bound is an *exact* and *rigorous* method, which guarantees an optimal solution or, if stopped before reaching one, returns the maximum error percentage in the best solution found. We think that Branch-and-Bound should also be considered, at least to tackle the smaller instances, as a viable option for these problems. Also, in the realm of local search heuristic procedures, there are several sophisticated paradigms that have proved effective for many hard combinatorial problems. In particular, Genetic Algorithms, Tabu Search with its variants, Simulated Annealing, Randomized Algorithms and various Metaheuristics. Given the relatively small effort required in implementing such searches (most of these procedures are available as general purpose libraries and the user has to implement only one or two core modules) we think that more work should be devoted into investigating which of these techniques is more suitable for a particular problem at hand.

RMSD

Given n points $(a^1, \ldots, a^n)$, with $a^i = (a_1^i, a_2^i, a_3^i)$ and n points $(b^1, \ldots, b^n)$, a "rigid body superposition", is a composition of a rotation and a traslation, that takes the points a^i into a'^i and makes the sum of differences $|a'^i - b^i|$ as small as possible.

More specifically, let Δ be defined by

$$\Delta^2 = \min_{R,t} \sum_i |Ra_i + t - b_i|,$$

where R is a rotation matrix (i.e., $\det R = 1$) and t a translation vector. The Root Mean Square Deviation of a and b is $\Delta/\sqrt{n}$.

The optimal traslation t is determined easily, since it can be proved that it must take the center of mass of the a^i points into the center of mass of the

b^is. To determine the optimal R one must consider the correlation matrix A defined by $A_{ij} = \sum_{h=1}^{n} a_i^h b_j^h$, as it can be shown that the optimal rotation maximizes $\text{Tr}(RA)$. One way to maximize $\text{Tr}(RA)$, exploits the fact that the problem is three dimensional. Represent the matrix R as

$$R = 1 + \sin\theta M + (1 - \cos\theta)M^2$$

where

$$M = \begin{pmatrix} 0 & n & -m \\ -n & 0 & l \\ m & -l & 1 \end{pmatrix}$$

and θ represent the rotation angle around an axis in the direction of the unit vector $u = (l, m, n)$. It then follows that

$$\text{Tr}(RA) = \text{Tr}(A) + \text{Tr}(MA)\sin\theta + (\text{Tr}(A) - \text{Tr}(M^2 A))\cos\theta$$

and hence

$$\max_{\theta} \text{Tr}(RA) = \text{Tr}(A) + \sqrt{(\text{Tr}(MA))^2 + (\text{Tr}(A) - \text{Tr}(M^2 A))^2}.$$

So, for a fixed axis of rotation, the calculation of the optimal angle θ is immediate [36].

Alternative ways of determining the optimal rotation matrix R, based on the computation of the eigenvalues of $A^T A$, or on the *quaternion* method, are described in [31].

As previously mentioned the RMSD is widely used as a sequence dependent measure of 3D similarity, i.e., to find an optimal mapping of points of a structure into a predetermined set of corresponding points in another structure. Despite some known flaws, the RMSD measure is at the hearth of many other measures. The main problem with RMSD is that it is more sensitive to small regions of local differences than to large regions of similarity. In fact, small distances between well-matched atoms have a much lesser impact on the RMSD than do the very large distances between poorly matched C_αs.

Another issue with RMSD is that it is not designed to be equivalent between different targets. For instance, the quality of a model with an RMSD of a 4 Å for a 40 residues long target is not the same as that of a model of 4 Å RMSD over 400 residues.

The problem of local different regions affecting the total score is common to other global measures, but for the RMSD value the effect is larger than for other scores, e.g., the contact map similarity. In this situation, a solution is typically to first calculate the similarity for segments of the protein and then define a normalized score based on the number of residues in the segments. However, the right relationship between the length of the segment and the value must be defined, and this is not a trivial problem. There are two conflicting issues: we are interested in large segments that are highly similar, but the similarity and the length of the segments are inversely proportional.

Similar problems are faced by "trimming" methods which, given a current alignment, re-define iteratively a core of pairs of residues that are matched within a small distance, work on these only, and then try to extend the alignment. The threshold and the degree of trimming are to some extent arbitrary, and this choice affects the final outcome.

DALI

In DALI [25], the structural alignment of two proteins is done indirectly, not by comparing the actual structures, but their *distance matrices*. The distance matrix of a folded protein P is the matrix $D^P = (d_{ij}^P)$ of the Euclidean distances between all pairs (i, j) of its residues. The matrix provides a 2D representation of a 3D structure, and contains enough information for retrieving the actual structure, except for overall chirality [22]. The idea underlying DALI is that if two structures are similar, then their distance matrices must be similar too. An analogous idea is used to compare structures via their *contact maps* and is described later in this article. In order to find similarities between two distance matrices, D^A and D^B, DALI uses a heuristic trick, and looks for all 6×6 submatrices of consecutive rows and columns in D^A and D^B (that is, it maps $i^A, i^A + 1, \ldots, i^A + 5$ into $i^B, i^B + 1, \ldots, i^B + 5$, and $j^A, j^A + 1, \ldots, j^A + 5$ into $j^B, j^B + 1, \ldots, j^B + 5$) to find patterns of similarity. This is done in the first step of the algorithm. The set $\{i^A, \ldots, i^A + 5, j^A, \ldots, j^A + 5\}$ with its counterpart in B, is called a *contact pattern*. The second step attempts at merging the contact patterns into larger, consistent, alignments. An alignment of two proteins A and B is a one-to-one function M of some residues M_A (the *matched* ones) of A into some residues of B. DALI tries to determine the alignment that maximizes the global objective function

$$S(M) = \sum_{i,j \in M_A} \phi(i, j, M(i), M(j)).$$

This objective function, which looks only at matched residues, depends on the particular ϕ used, which in turn depends on the C_α–C_α distances d_{ij}^A and $d_{M(i),M(j)}^B$. The two main forms for ϕ used by DALI are a *rigid* and an *elastic* score. The rigid score is defined by $\phi(i, j, M(i), M(j)) := \theta - |d_{ij}^A - d_{M(i),M(j)}^B|$, where $\theta = 1.5$Å is the zero level of similarity. The more complex elastic score uses relative, other than absolute deviations, and an envelope function to weigh down the contribution of long distance matched pairs.

Some heuristic rules are employed to speed up the search. For instance, only a subset of all contact patterns is considered. Also, overlapping contact patterns (they could possibly overlap by 11 out of 12 residues) are suppressed by partitioning the protein in nonoverlapping structural elements and merging repetitive segments. Further rules to prune the search involve the suppression of pairs of patterns for which the sum of distances of one is not within a given interval of the same value in the other. Other rules of the same nature are employed which we do not describe here. The final optimization is done via a Monte Carlo algorithm, which is basically the local search strategy known as Simulated Annealing. In this search a *neighborhood* of a solution is explored for a good *move*, leading to a new solution. A move which takes into a solution worse than the current one can be accepted, but with probability inverse to the degradation in the solution quality. The *temperature* of the system is also used to favour or make more difficult such non–improving moves. DALI uses two basic moves, named *expansion* and *trimming* ones. Expansion moves try to extend a current partial alignment by adding to it a contact pattern compatible with one of the currently matched residues (i.e., containing the same match), and then possibly removing some matches that have become noncompatible. Trimming moves simply remove from a current alignment any subalignment of 4 elements that gives negative contribution to the total similarity score. These moves are alternated as one trimming cycle every five expansion cycles.

DALI was used to carry out an all–against–all structure comparison for 225 representative protein structures from the PDB, providing the basis for classification in the FSSP database. Also, the DALI server can be used to submit the coordinates of a query protein structure which is then compared against the PDB.

MaxSub

MaxSub [50] is an algorithm explicitly developed to be used for the automatic assessment of protein structure similarity (in particular within the CAFASP exercise). Therefore it was written with the goals of (a) being simple and objective, (b) producing one single number (score) that measures the amount of similarity between two structures and (c) performing similarly to human-expert evaluations. These goals were partly met, and the measure was partially validated when, on a particular category and subset of predictions, it ranked in the top the same groups that were ranked in the top by the CASP human evaluators. A prediction can be evaluated in different aspects. For example, in the Homology Modeling category one may want to score the accuracy of the loops or the side chain packing. Or, in the Fold Recognition category, one may want to evaluate wether a compatible fold was recognized, regardless of the quality of the alignment obtained. MaxSub, in particular, focuses on the quality of the alignment of the models to the target. It is a *sequence–dependent* assessment, i.e., only corresponding residues are compared. The result is a single scalar in the range of 0 (completely wrong) to 1 (perfect model). The scalar is a normalization of the size of the largest "well–predicted" subset, according to a variation of a formula devised by Levitt and Gerstein [32]. MaxSub is also similar to GDT, a method developed by Zemla et al. [59] for being used in CASP3, which attempts to find a large set of residues which can be superimposed over the experimental structure within a certain error. GDT is briefly described later in this survey.

The input to MaxSub are the 3D coordinates $A = \{a_1, \ldots, a_n\}$ and $B = \{b_1, \ldots, b_n\}$ of the C_α atoms of a prediction and a target. A solution is a subset $M \subset \{1, \ldots, n\}$ (called a *match*) such that the pairs (a_j, b_j) for $j \in M$ can be superimposed "well-enough" (i.e., below a given threshold d) by a transformation T (rotation and translation). The objective is to maximize the size of M, i.e., find the largest subset of residues which superimpose well upon their correspondents.

Given M, the value of the best superposition can be easily found, (analogously to the RMSD determination previously described), by using the transformation T_M which minimizes, over all T,

$$RMS(M) = \sqrt{\frac{\sum_{j \in M} ||a_j - T(b_j)||}{|M|}}$$

where $|| \cdot ||$ is the Euclidean norm.

The optimization of M is performed by a simple heuristic algorithm. This algorithm is based on the reasonable (although not rigorous) assumption that a good match must contain at least $L \geq 4$ consecutive pairs, i.e. $\{(a_i, b_i),\ i = j, j+1, \ldots, j+L-1\}$ for some j. So the idea is, for a fixed L, to look, in time $O(n)$, for all matches of L consecutive elements, and try to extend each of them. The extension of a match M is also done heuristically, in 4 iterations, where at each iteration k the pairs (a_i, b_i) for which $||a_i - T_M(b_i)|| < kd/4$ are added to M.

Let M^* be the best match found by the above procedure. In order to return a single, normalized score, a final value S is computed as

$$S(M^*) = \left(\sum_{i \in M^*} \frac{1}{1 + (\frac{d_i}{d})^2} \right) / n,$$

where $d_i = ||a_i - T_{M^*}(b_i)||$. In our opinion, there are several non rigorous steps that should be addressed in evaluating MaxSub and its reliability. For instance, although the basic objective optimized by the algorithm is *maximize* $|M|$ *such that* $d_i \leq d$ *for all* $i \in M$, the optimization of S (which is the score eventually returned) calls ultimately for *maximize over all* M, $\sum_{i \in M} \frac{1}{d^2 + d_i^2}$ which is a different objective with possibly a different optimum. Also, there are many arbitrary parameters, such as L, the number 4 of iterations in the extension phase, and the threshold d. Some experiments have shown that the algorithm does not depend on the choice of d too much, while the dependance on the other parameters is not described with the same accuracy. As far as how close the solution is to the optimum, a comparison to the best of a random set of 70,000 matches was used. Given the astronomical number of possible solutions, it is arguable if such method can be used to conclude anything about the actual performance of the algorithm.

Lgscore

The measure Lgscore [9] is used in LiveBench and CAFASP for the automatic assessment of fold recognition problems. This measure is statistically based, and relies on the following formula, by Levitt and Gerstein [32], for the similarity of two structures, after a superposition in which M denotes the aligned residues:

$$S_{\text{str}}(M) = K \left(\sum_{i \in M} \frac{1}{1 + (d_i/d_0)^2} - \frac{N_g}{2} \right),$$

where K and d_0 are constants (usually set to 10 and 5Å), d_i is the distance between the aligned pairs i of C_α atoms, and N_g is the number of gaps in the alignment. The P–value of a similarity is the likelihood that such a similarity could be achieved by chance. Levitt and Gerstein showed how to compute the P–values for a distribution of S_{str} depending on the length of the alignment. Lgscore is the negative log of the P–value for the most significant subpart of the alignment. In order to find such most significant segment, two heuristic algorithms are used, i.e. "top–down" and "bottom–up". The top–down approach consists of a loop in which (1) a superposition is done of all residues that exist in the current model and the target; (2) the P–value of this superposition is stored and the residues that are furthest apart are deleted in the model and the target. The loop is repeated as long as there are at least still 25 residues. The alignment with the best P–value is returned. The bottom–up approach essentially tries to match a fragment $i, \ldots, i+j$ of $j \geq 25$ consecutive residues in the model and the target, for different values of i and j and returns the best P–valued fragment. None of the two approaches dominates the other, although the authors report that the bottom–up algorithm found the best subset in most cases. The arbitrary value of 25 residues as a threshold for fragment length is justified in [9] since "short segments are given unrealistic good P–values".

GDT

GDT (Global Distance Test, [59]) is an algorithm for identifying large sets of residues (not necessarily continuous) in a prediction and a target which do not deviate more than a given cutoff value. The algorithm is heuristic, and similar to Lgscore. It consists of a loop, started from the alignment between model and target of the longest continuous fragments within the cutoff value, found by enumeration, and all three–residue segments. At each iteration, an RMS transfom is computed for the given alignment, and the pairs of residues whose distance is greater than the threshold are removed. The loop is ended as soon as no residues are removed by one iteration.

Geometric Hashing and 3dSEARCH

Comparing protein structures and, more generally, querying databases of 3D objects, such as the PDB, can be regarded as special cases of Object Recognition problems in Computer Vision. Therefore, algorithms originally developed in the field of Computer Vision have now found a suitable application in Structural Genomics as well. *Geometric hashing* [28] is an example of such a technique. Geometric Hashing can be used to match a 3D object (a protein structure) against one or more similar objects, called "models" (e.g. a database organized on structure). The key idea is to represent each model under different systems of coordinates, called *reference frames*, one for each triplet of non collinear points of each model. In any such system, all points of the model (which are vectors of reals) can be used as keys (i.e., indices in a hash, or look–up, table) to store the information about the model they belong to. In a *preprocessing phase*, all models are processed this way, and an auxiliary, highly redundant, hash table is created. In the *querying phase*, executed each time a target is to be matched against the models, the following steps are taken. For each reference frame of the target, the points of the target (in the coordinate system defined by the reference frame) are used to access the hash table and hence retrieve some of the models. These models define a *match list* for the particular reference frame. A match list gives a set of alignments of the points in the target and in each of the models retrieved. The match lists are then merged in order to find larger alignments. The final step is to compute an RMSD transformation for each alignment obtained, so as to find the model most similar to the target.

The program 3dSEARCH [51], used by the database SCOP, implements the strategy of geometric hashing on the secondary structure representation of each protein.

MNYFIT

The program MNYFIT [54], which is used by the database HOMSTRAD, belongs to a suite of programs called COMPOSER, for homology modeling of protein structures. With COMPOSER one can work out the construction of a predicted 3D structure based on one or more known homologous ones. MNYFIT is a core module within this software collection, and can handle up to 10 molecules of about 500 residues each. It performs the superposition of two or more related protein structures, individues structurally equivalent residues and their location, and with this information is able to define a common framework for a set (family) of proteins.

The main algorithm used by MNYFIT is the least squares fitting procedure of McLaughlan [36] for rigid body superposition, akin to the RSMD procedure discussed above.

The superpostion is performed initially using at least three atoms from the C_α backbones of each molecule, which occupy topologically equivalent positions in all the molecules to be fitted. For two or more superimposed structures, a threshold is used to define which of the C_α atoms are considered structurally equivalent (i.e. those whose distance in the superposition is smaller than the threshold). Then, the alignment is iteratively updated with the double objective of increasing the number of structurally equivalent atoms while at the same time keeping a low Root Mean Square Deviation between the equivalenced atoms in the superimposed structures. When more than two molecules are superimposed, the objective is to determine a consensus structure (called "framework"), which should be as close as possible to all involved structures, and which represents the average positions of the structurally equivalent atoms of all superimposed molecules. The framework returned from the final step of the procedure is used by COMPOSER for further model–building operations.

STAMP

One of the programs used by the database HOMSTRAD for the detection of homologous protein families is STAMP [48], a program for the alignment of protein sequences based on their 3D structures. The core of this program contains an implementation the basic Smith and Waterman dynamic programming procedure for sequence alignment [52], but using suitable similarity scores which express the probability of residue–residue structural equivalence. These scores are computed according to the equation by Argos and Rossmann [1]

$$P_{ij} = \exp \frac{d_{ij}^2}{-2E_1^2} \exp \frac{s_{ij}^2}{-2E_2^2},$$

where d_{ij} is the distance between the C_α atoms of residues i and j, and s_{ij} measures the local main chain conformation. The Smith and Waterman procedure is embedded in a loop as follows: at each iteration, the output of the dynamic program is a residue–residue equivalence (the sequence alignment). This list of equivalences is used to compute a best fit transformation minimizing the RMSD, via the least square method of McLaughlan. The new set of coordinates and residue–residue distances, obtained under this transformation, are then used to recompute the similarity score values, which are then used for another round of the Smith Waterman procedure. The loop is repeated iteratively until the alignment becomes stable.

The method has proved experimentally effective, allowing the generation of tertiary structure based multiple protein sequence alignments for a variety of protein structural families. However, the method is only effective for proteins which share a good deal of global topological similarity, while fails if applied to, e.g., proteins with common secondary structures but with different connectivity, orientation or organization.

STAMP allows for the specification of a minimum number of equivalent residues to be matched in two structures, the reversals of strand directions, the swapping of sequence segments and more. The output contains two measures of alignment confidence: a "structural similarity score", which can also be used to measure the functional and evolutionary relationship, and an "individual reside accuracy" which is intended to measure the quality of the topological equivalence of the pairs of aligned residues.

5 Problems Based on Contact Map Representations

A *contact map* [33, 21] is a binary version of the distance matrix representation of a protein structure. More specifically, the contact map of a folded protein of n residues is a 0-1, $n \times n$ matrix C, whose 1–elements correspond to pairs of amino acids in three–dimensional "contact". A contact can be defined in many ways. Typically [37], one considers $C_{ij} = 1$ when the distance of two heavy atoms, one from the i–th aminoacid and one from the j–th aminoacid of the protein, is smaller than a given threshold (e.g., 5Å). The framework of the contact map representation of proteins is very appealing, since this intuitive and fairly simple representation is already complex enough to capture the most important properties of the folding phenomenon. It has been shown that it is relatively easy to go from a map to a set of possible structures to which it may correspond [21, 56]. This result has opened the possibility of using contact maps to predict protein structure from sequence, by predicting contact maps from sequence instead. Vendruscolo and collaborators, among others, have looked at the problem of devising an energy function based on contacts, which should be minimized by the protein's native state contact map [57, 45]. For this purpose, they have set up a system of linear inequalities, with 20×20 variables C_{ab} for all pairs of aminoacids a and b, which represent the weight to give to a contact between the aminoacids a and b. The inequalities are built as follows. Given the contact map of a correct structure r, there is an inequality for any alternative structure w over the same sequence of aminoacids, imposing that the energy of r is lower than that of w. Alternative structures are obtained, e.g., by threading through other known folds. The results are that "a simple pairwise contact energy function is not capable of assigning the lowest energy to an experimentally determined structure" [45], but by using corrective factors, such as a distinction between contacts on the surface or in the core, and simple distance–dependant interaction weights, one can achieve contact potentials which are in fact often stabilized by the native contact maps. The use of energy function minimization to predict contact maps is just one possible way. To the best of our knowledge, very little success has been met so far in the contact map prediction problem. It is possible that the research on this question will be boosted by the fact that the competition CAFAPS has recently introduced the new "contacts" prediction category.

The statistics of contact maps have been studied as well, and it has been shown that the number of contact maps corresponding to the possible configurations of a polypeptide chain of n residues, represented by a self–avoiding

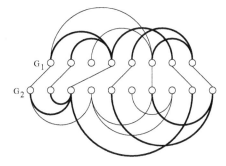

Fig. 2. An alignment of value 5

walk in the d–dimensional lattice, grows exponentially with n for all $d \geq 2$ [58].

5.1 Using Contact Maps for Structure Comparison

Besides their use for protein fold prediction, contact maps can be exploited to compare 3D structures. The basic idea is fairly obvious: if two structures are similar, we should expect their contact maps to be similar as well, and conversely. Hence, we can use an indirect method for structure comparison, i.e., contact map comparison instead. In our previous work [29] we have designed an exact algorithm based on an Integer Programming formulation of this problem, which we will now review.

We can regard the contact map of a protein p as the adjacency matrix of a graph G_p. Each residue is a node of G_p, and there is an edge between two nodes if the the corresponding residues are in contact. The *Contact Map Overlap* (CMO) problem, calls for determining a sequence–independent alignment of some residues in the first protein (nodes in G_1) with residues of the second protein (nodes in G_2) which highlights the largest set of common contacts as follows: The value of the alignment is the number of contacts (i.e., edges) in the first map whose endpoints are aligned with residues that are also in contact in the second map. This value is called the *overlap* for the two proteins, and the optimization problem is to find the maximum overlap. Figure 5.1 shows two contact maps and a feasible alignment.

This measure was introduced in [15], and its optimization was proved NP-hard in [18], thus justifying the use of sophisticated heuristics or Branch–and–Bound methods.

Some of the most powerful algorithms for finding exact solutions of combinatorial optimization problems are based on *Integer Programming* (IP) [43]. The IP approach consists in formulating a problem as the maximization of a linear function of some integer variables and then solving it via Branch–and–Bound. The upper bound comes from the *Linear Programming* (LP) *relaxation*, in which the variables are not restricted to be integer, and is polynomially solvable. When the LP–relaxation value is close to the value over the integers, then the bound, and hence the pruning of the search space, is effective. In order to obtain good bounds, the formulation is often reinforced by the use of additional constraints, called *cuts* (from which the approach name, *Branch–and–Cut*): these are constraints that do not eliminate any feasible integer solution, but make the space of fractional solutions smaller, this way decreasing the value of the LP bound. In many cases a good IP formulation requires an exponential number of constraints and/or cuts. This would make its practical solution impossible, unless there is a way to include all of them only implicitly. This way exists, and works as follows. Given an LP fractional solution x^* and an inequality $ax \leq b$, we say that the inequality is *violated* by x^* if $ax^* > b$. If we have an exponential number of inequalities, we can solve the LP with only a (small) subset of them, obtain a solution x^* and then check if any of the (exponentially many) inequalities that were left out is violated by x^*. If not, the current solution is optimal with respect to the full formulation, otherwise, we can add the violated inequality to the current LP and iterate the process. The check for a violated inequality is called *separation* and is carried out by a *separation algorithm*. By a fundamental result of Grötschel, Lovász and Schrijver [20], the existence of a polynomial–time separation algorithm is a necessary and sufficient condition for solving the whole exponential–sized LP relaxation in polynomial time.

The CMO problem can be reduced to a (very large) *Maximum Independent Set* (MIS) problem on a suitable graph. An *independent set* is a set of vertices such that there is no edge between any two of them. The MIS is a classic problem in combinatorial optimization which has a definition nice and simple, but is one of the toughest to solve exactly. However, by exploiting the particular characteristics of the graphs derived from the CMO problem, we can in fact solve the MIS on graphs of 10,000 nodes and more.

The natural formulation of the MIS as an IP problem has a binary variable x_v for each vertex v, with the objective of maximizing $\sum_v x_v$, subject to $x_u + x_v \leq 1$ for all edges $\{u, v\}$. This formulation gives a very weak bound, but it can be strengthened by the use of *clique inequalities* cuts, such as $\sum_{v \in Q} x_v \leq 1$, which say that any clique Q can have at most one node in common with any independent set. The addition of these constraints can lead to tight formulations. This is exactly the case for the CMO problem. In [29] we formulated the CMO problem as an IP and solved it by Branch–and–Cut, where the cuts used are mainly clique–inequalities. Although there is an exponential number of different clique inequalities, we can characterize them completely and separate over them in fast $(O(n^2))$ polynomial time. Note that finding cliques in a graph is in general a difficult problem, but in this case we can solve it effectively since it can be shown that the underlying graph is perfect. The following section gives some details on the formulations and results from [29]. Further details and full proofs can be found in the original paper.

IP Formulation

We can phrase the CMO problem in graph–theoretic language as follows: We are given two undirected graphs $G_1 = (V_1, E_1)$ and $G_2 = (V_2, E_2)$, with $n_i = |V_i|$ for $i = 1, 2$. A total order is defined on $V_1 = \{a_1 < \ldots < a_{n_1}\}$ and $V_2 = \{b_1 < \ldots < b_{n_2}\}$. It is customary to draw such a graph with the vertices arranged increasingly on a line. We denote an edge by an ordered pair (i, j), with a tail in the left endpoint and a head in the right endpoint.

A non–crossing alignment of V_1 in V_2 is defined by any two subsets of the same size k, $\{i_1, \ldots, i_k\} \subseteq V_1$ and $\{u_1, \ldots, u_k\} \subseteq V_2$, where $i_1 < i_2 \ldots < i_k$ and similarly for the u_h's. In this alignment, u_h is aligned with i_h for $1 \leq h \leq k$. Two edges (contacts) $(i, j) \in E_1$ and $(u, v) \in E_2$ are *shared* by the alignment if there are $l, t \leq k$ s.t. $i = i_l$, $j = i_t$, $u = u_l$ and $v = u_t$ (see Figure 5.1). Each pair of shared edges contributes a *sharing* to the objective function. The problem consists in finding the non–crossing alignment which maximizes the number of sharings.

An alignment corresponds to a set of lines connecting nodes of V_1 and V_2 in the usual drawing with V_1 drawn on the top and V_2 on the bottom. We denote such a line for $i \in V_1$ and $j \in V_2$ by $[i, j]$. We say that two lines *cross* if their intersection is a point. The sharings (e_1, f_1) and (e_2, f_2) can be both achieved by an alignment if and only if they are *compatible*, i.e. no two of the lines betweens the tails of e_1 and f_1, the tails of e_2 and f_2, the heads of e_1 and f_1 and the heads of e_2 and f_2 cross. A set of sharings is *feasible* if the sharings are all mutually compatible, otherwise it is *infeasible*. Similarly we define a feasible and infeasible set of lines. If we draw the lines connecting the endpoints of an infeasible set of sharings, we have an infeasible set of lines.

We denote by y_{ef} a binary variable for $e \in E_1$ and $f \in E_2$, which is 1 iff the edges e and f are a sharing in a feasible solution. The objective function of CMO is

$$\max \sum_{e \in E_1, f \in E_2} y_{ef}, \qquad (1)$$

and the constraints are

$$y_{e_1 f_1} + y_{e_2 f_2} \leq 1, \qquad (2)$$

for all $e_1, e_2 \in E_1$, $f_1, f_2 \in E_2$ s.t. (e_1, f_1) and (e_2, f_2) are not compatible.

It can be shown that the formulation made of constraints (2) gives a very weak LP bound, and also contains too many constraints. To strengthen the bound, we use a new set of variables and strong cuts. We introduce a new set of binary variables x_{iu} for $i \in V_1$ and $u \in V_2$, and constraints forcing the nonzero x variables to represent a non–crossing alignment. We then bound the y variables by means of the x variables, so that the edges (i, j) and (u, v) can be shared only if i is mapped to u and j to v. For $i \in V_1$ (and analogously for $i \in V_2$), let $\delta^+(i) = \{j \in i + 1, \ldots, n_1 : (i, j) \in E_1\}$ and $\delta^-(i) = \{j \in 1, \ldots, i - 1 : (j, i) \in E_1\}$. Then we have the following constraints:

$$\sum_{j \in \delta^+(i)} y_{(i,j)(u,v)} \leq x_{iu}, \qquad \sum_{j \in \delta^-(i)} y_{(j,i)(u,v)} \leq x_{iv} \qquad (3)$$

for all $i \in V_1$, $(u,v) \in E_2$, and analogous constraints for $i \in V_2$ and $(u,v) \in E_1$. We call these *activation* constraints. Finally, the *noncrossing* constraints are of the form:

$$x_{iu} + x_{jv} \leq 1 \qquad (4)$$

for all $1 \leq i \leq j \leq n_1$, $1 \leq v \leq u \leq n_2$ s.t. $i \neq j \lor u \neq v$.

Our IP formulation for the max CMO problem is given by (1), (3), and (4), where x and y are all binary variables. The cuts in the x variables are described next.

First, we make clear the connection to the independent set problem. We define two graphs G_x and G_y. In G_x there is a node N_{iu} for each line $[i,u]$ with $i \in V_1$ and $u \in V_2$ and two nodes N_{iu} and N_{jv} are connected by an edge iff $[i,u]$ and $[j,v]$ cross. Similarly, in G_y there is a node N_{ef} for each $e \in E_1$ and $f \in E_2$ and two nodes N_{ef} and $N_{e'f'}$ are connected by an edge iff the sharings (e,f) and (e',f') are not compatible. Then, a selection of x variables feasible for all noncrossing constraints corresponds to an independent set in G_x and a feasible set of sharings is an independent set in G_y. The maximum independent set in G_y is the optimal solution to CMO. All cuts valid for the independent set problem can be applied to the x and y variables, and, most notably, the clique inequalities.

The separation of the clique inequalities works as follows. Given a fractional solution x^*, we look for the *maximum weight clique* in G_x, where x_{ij}^* is the weight of N_{ij}. If the maximum weight clique weighs less than 1, then there are no violated clique inequalities, otherwise, we can add at least one such cut to the current LP.

Finding maximum weight cliques is a polynomial problem when the underlying graph is perfect. We can prove that

Theorem 5.1. *The graph G_x is perfect.*

There are algorithms for finding a max weighted clique in a weakly triangulated graph (as G_x can be shown to be) of time $O(n^5)$, due to Hayward, Hoang, Maffray [23] and Raghunathan [46]. However, for this specific graph we can do better and find max weighted cliques in time $O(n^2)$, thus making a huge difference in the practical solution of the problem.

We start by chacterizing all cliques in G_x, i.e. sets of alignment lines which are all mutually crossing. We define the following notion of a *triangle*: this is a set of lines with one common endpoint and a second endpoint in a range of consecutive nodes, like $T(i,j|u) := \{[i,u], [i+1,u], \ldots, [j-1,u], [j,u]\}$ where $i \leq j \in V_1$ and $u \in V_2$, and $T(i|j,u) := \{[i,j], [i,j+1], \ldots, [i,u-1], [i,u]\}$ where $i \in V_1$ and $j \leq u \in V_2$. For S a set of lines, by $x(S)$ we denote the value $\sum_{[i,j] \in S} x_{ij}$.

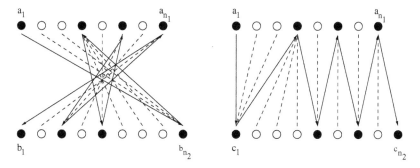

Fig. 3. Left: A zigzag path P (bold) and the set $T(P)$. Right: Same path after flipping $V2$.

Call a_1, a_{n_1}, b_1, and b_{n_2} the set of *terminal* nodes. Consider a path P which passes through all the terminal nodes, and alternates nodes of V_1 and V_2 in a zig–zag fashion: That is, we can orient the path so that a_1 is the first of the nodes of V_1 visited by the path, and if a_k has been visited by the path, then all of the nodes in V_1 visited after a_k are to its right. Similarly, b_{n_2} is the first of the nodes of V_2 visited by the path, and if b_h has been visited by the path, then all of the nodes in V_2 visited after b_h are to the left. Note that any such path must start and end at a terminal node (see figure 5.1, Left), and must always include the lines $[a_1, b_{n_2}]$ and $[a_{n_1}, b_1]$. For each node of degree two in P, a triangle is defined by considering the set of lines incident on the node and contained within the two lines of the path. Let $T_A(P)$ $(T_B(P))$ be the set of triangles defined by P with tip in the nodes of V_1 (V_2) having degree two in P. We define $T(P) := T_A(P) \cup T_B(P)$. The following theorem characterizes the cliques of G_x.

Theorem 5.2. *A set Q of lines is a maximal clique in G_x if and only if there exists a zigzag path P such that $Q = T(P)$.*

The inequalities $x(T(P)) \leq 1$ for all zigzag paths P are therefore the strongest clique cuts for this particular independent set problem. We now show that to find the most violated such inequality in time $O(n^2)$. To make the following argument easier, we rename the nodes of V_2 as $\{c_1, \ldots, c_{n_2}\}$, so that the leftmost node c_1 is b_{n_2} and the rightmost, c_{n_2}, is b_1 (i.e., we flip the nodes of V_2 with respect to the usual drawing). A zigzag path P now looks as a path which goes from left to right both in V_1 and V_2. We call such a path a *leftright* path (see figure 5.1, Right).

With respect to the new drawing of W, orient each line $[a, c]$ in the two possible ways and, given a real vector x^*, define the length for each arc $(a, c) \in V_1 \times V_2$ and $(c, a) \in V_2 \times V_1$ as follows: $l(a, c) = x^*(T(a|1, c)) - x^*(T(1, a|c))$ and $l(c, a) = x^*(T(1, a|c)) - x^*(T(a|1, c))$. The lengths of four special arcs are defined separately, as $l(a_1, c_1) = 0$, $l(c_1, a_1) = 0$, $l(a_{n_1}, c_{n_2}) = x^*(T(a_{n_1}|1, c_{n_2}))$

and $l(c_{n_2}, a_{n_1}) = x^*(T(1, a_{n_1}|c_{n_2}))$. Now, consider a leftright path P starting with either the arc (a_1, c_1) or (c_1, a_1) and ending with either the arc (a_{n_1}, c_{n_2}) or (c_{n_2}, a_{n_1}). Call $l(P)$ the standard length of this path, i.e. the sum of arcs lengths. We then have the following lemma.

Lemma 5.1. *For a leftright path P, $l(P) = x^*(T(P))$.*

Hence, to find the max–weight clique in G_x, we just need to find the longest leftright path. This can be computed effectively, by using Dynamic Programming. Call $V_\searrow(i, j)$ the length of a longest leftright path starting at a_i and using nodes of V_2 only within $c_j, c_{j+1}, \ldots, c_{n_2}$. Also, call $V_\nearrow(i, j)$ the length of a longest zigzag path starting at c_j and using nodes of V_1 only within $a_i, a_{i+1}, \ldots, a_{n_1}$. Then we have the following recurrences:

$$V_\searrow(i, j) = \max\{l(a_i, c_j) + V_\nearrow(i+1, j), V_\searrow(i, j+1)\},$$

$$V_\nearrow(i, j) = \max\{l(c_j, a_i) + V_\searrow(i, j+1), V_\nearrow(i+1, j)\}.$$

These recurrences can be then solved backwards, starting at (n_1, n_2), in time $O(n^2)$.

Genetic Algorithms and Local Search

In a Branch–and–Cut algorithm, pruning of the search space happens whenever the upper bound ub to the optimal solution of a subroblem is smaller than a lower bound lb to the global optimum. As described before, ub is the value of the LP–relaxation. It is important then to have good (i.e., large) lower bounds as well. Any feasible solution gives a lower bound lb to the optimum. In our work we have developed heuristics of two types for generating good feasible solutions, i.e. *Genetic Algorithms* and *Steepest Ascent Local Search*.

A generic Genetic Algorithm (GA, [24, 16]) mimics an evolutionary process with a population of individuals that, by the process of mutation and recombination, gradually improve over time. For optimization problems, an individual is a candidate solution, mutation is a slight perturbation of the solution parameters, and recombination is a "merging" of two candidate solutions. The use of GAs for Structural Genomics problems is not new, e.g., GAs were used for protein fold prediction [27, 10].

Our application of the GA encodes the solutions as sets of alignment edges, associating a residue in one contact map graph with a residue in the other. A mutation will slightly shift one edge, and randomly add new edges in any available space, while recombination will pick edges out of two candidate solutions and create a new candidate solution using those edges.

Figure 5.1 shows two mutations. The first mutation shifts the alignment edges on the circled nodes to the right by one position, causing the right-most edge to be removed and a new edge to be be inserted. The second mutation shifts the circled edges to the left by one position, causing the left-most shifted

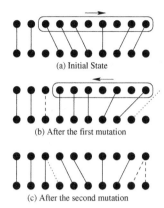

(a) Initial State

(b) After the first mutation

(c) After the second mutation

Fig. 4. The mutation operator. The top alignment shows the state before any mutations, the middle alignment shows the state after the first mutation but before the second mutation, the bottom alignment shows the state after both mutations. Dotted lines are edges that have been removed by a mutation, dashed lines are edges that have been added after a mutation is performed.

edge to be removed and a new edge to be randomly inserted on the right end – exactly one of the dashed lines is inserted.

The recombination operator is used to create a new candidate solutions (the child) from existing solutions. This is done by randomly selecting two existing solutions (the parents) by a standard GA method, i.e., randomly, but biased towards good solutions. A set of contiguous edges is randomly selected from one of the parents and is copied directly to the child. Next, all edges from the second parent that do not cross edges in the child are copied to the child. Finally, new edges are added in any available positions, exactly as was done with mutation.

Our Steepest Ascent Local Search heuristic algorithms follow the standard approach: Let s be a feasible current solution. The *neighborhood of s* is the set of solutions which can be obtained by applying a *move* to s. If all solutions in the neighborhood are no better than the current solution s, s is a local optimum and the search is terminated. Otherwise, the move that results in the best solution value is applied to s, and the search continues. Since converging to a local optimum is very fast, the search can be repeated many times, each time starting from a random feasible solution.

A feasible contact map alignment solution is identified by a pair of lists of the same size, (A, B), where $A = (a_1 < \ldots < a_k)$ are nodes from G_1 and $B = (b_1 < \ldots < b_k)$ are nodes from G_2. The alignment maps residue a_i in the first graph to residue b_i in the second graph (see Figure 5.1(a)).

Our two local search algorithms differ only in the definition of the moves that generate a neighborhood. The first algorithm uses moves that add a single specific line to the solution, removing any lines that cross the new line.

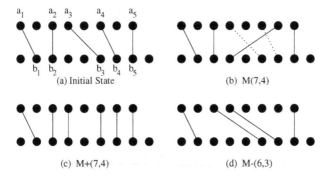

(a) Initial State

(b) M(7,4)

(c) M+(7,4)

(d) M-(6,3)

Fig. 5. Moves for the two algorithms. (a) Starting solution s. (b) A move from the first algorithm applied to s. (c) An increasing move applied to s. (d) A decreasing move applied to s.

Formally, the move $M(a, b)$ is defined for all $a \in G_1 - A$ and $b \in G_2 - B$ and will add the line $[a, b]$, removing all lines $[a_j, b_j]$ such that $a < a_j \wedge b > b_j$ or $a > a_j \wedge b < b_j$ (see Figure 5.1(b)). This moves allows for big "jumps" in the solution space, by introducing very skewed lines and by removing many lines at once, and is suitable for instances in which the contact maps are not similar.

The second algorithm uses two types of moves, a *decreasing* move and an *increasing* move. The decreasing move, $M^-(a, b)$, defined for $a \in A$ and $b \in B$, simply removes a from A and b from B. Figure 5.1(d) shows the decreasing move $M^-(6, 3)$, which removes a_4 and b_2 from their respective lists. The increasing move, $M^+(a, b)$, where a and b are as defined for $M(a, b)$ in the first algorithm, adds a to A and b to B (see Figure 5.1(c)). These moves do not introduce very skewed lines easily and so are suited for similar proteins, in which good solutions are made of many parallel lines.

Computational Experiments

Our program has been implemented and run on some proteins from the PDB. This is the first time that exact solutions have been found for real instances of this problem. We have run our procedure on a set of 269 proteins, with sizes ranging from 64 to 72 residues and 80 to 140 contact each. The set was chosen to contain both a large number of similar proteins, as well as a large number of dissimilar proteins. An all-against-all computation would have resulted in 36046 alignments; we selected a subset of 597 alignments, so that there would be an equal number of similar pairs and dissimilar pairs.

In order to perform such a massive computation on a standard, single–processor Pentium PC, we have limited the time devoted to each individual instance to one hour (details can be found in [29]). Therefore, some problems have not been solved to optimality. However, even within the time limit

	Fold	Family	Residues	Seq. Sim.	RMSD	Proteins
1	Flavodoxin-like	CheY-related	124	15-30%	< 3Å	1b00, 1dbw, 1nat, 1ntr, 1qmp, 1rnl, 3cah, 4tmy
2	Cupredoxins	Plastocyanin/ azurin-like	99	35-90%	< 2Å	1baw, 1byo, 1kdi, 1nin, 1pla, 2b3i, 2pcy, 2plt
3	TIM beta/ alpha-barrel	Triosephosphate isomerase	250	30-90%	< 2Å	1amk, 1aw2, 1b9b, 1btm, 1hti, 1tmh, 1tre, 1tri, 1ydv, 3ypi, 8tim
4	Ferratin-like	Ferritin	170	7-70%	< 4Å	1b71, 1bcf, 1dps, 1fha 1ier, 1rcd

Fig. 6. The Skolnick set.

set, we have been able to solve 55 problems optimally and for 421 problems (70 percent) the gap between the best solution found and the current upper bound was less than or equal to 5, thereby providing a strong certificate of near–optimality. The results also show the effectiveness of our lower bounding heuristic procedures, and in particular, of our genetic algorithm. The GA heuristic turned out to be superior to the others, finding 52 of the 55 optimal solutions.

In a second test, we used our programs to cluster proteins according to their Contact Map Overlap. I.e., given a set of proteins, we compute a normalized score based on their best CMO alignment, and check if pairs with high score are actually coming from the same family.

5.2 The Skolnick Clustering Test

A test set was suggested to us by Jeffrey Skolnick. The set contains 33 proteins classified by SCOP into four families: CheY-related, Plasto–cyanin/azurin-like, Triosephosphate isomerase and Ferratin (see Figure 5.1). Since this proteins are relatively large (beyond the capability of Branch–and–Cut exact solution, the problem is NP–hard after all), we used the heuristics that worked well for providing lower bounds in the Branch–and–Cut algorithm, i.e. GA and Local Search.

We applied the heuristics to all 528 contact map pairs, and clustered the proteinss based on the following similarity score:

$$s_{ij} = \frac{c_{ij}}{\min\{m_i, m_j\}}$$

where m_i and m_j are the number of contacts in proteins i and j, and c_{ij} is the number of shared contacts in the solution found. Note that $0 \le s_{ij} \le 1$. A score of 1.0 would indicate that the smaller contact map is completely contained in the larger contact map. We found that 409 alignments with score between 0.0 and 0.313 were almost exclusively between contact maps in different families; 1.3% of these were alignments within the same family. The remaining 119 alignments, with score between 0.314 and 0.999, were all between contact maps in the same family. Hence, we validated the CMO score for clustering with 98.7% accuracy (1.3% false negatives).

5.3 Conclusions

In order to promote the use of contact map overlap for structure comparison, more work should be devoted into designing effective algorithms for proteins of larger size than our current limit. Armed with such an algorithm, one could run an all against all alignment of the PDB structures, and cluster them according to their contact map similarity.

Our Branch–and–Bound and heuristic algorithms can be extended to new contact map–based similarity measures. For instance, one can consider the introduction of weights w_e on the contacts $e = \{i_1, i_2\}$. These weights can be based on the residues appearing at positions i_1 and i_2, or on the distance between i_1 and i_2 in the folded protein. This would allow to model a situation in which some contacts are "more important" to preserve than others. Similarly, penalties u_{ij} can be used to weigh each residue–residue x_{ij} alignment. Finally, the objective function could be made parametric in the number of residues considered, so that the measure can be used, e.g., for local alignments. The problem would then read: for a given k find the best contact map alignment which maps exactly k residues of the first proteins in k of the second.

These and similar measures will be the object of our future research.

6 Acknowledgements

Support for Lancia was provided in part by the Italian Ministry of University and Research under the National Project "Bioinformatics and Genomics Research", and by the Research Program of the University of Padova.

References

1. P. Argos and M. Rossmann. Exploring structural homology of proteins. J. Mol. Biol. **105** (1976) 75–95
2. F. Allen *et al.* Blue Gene: A vision for protein science using a petaflop super-computer. IBM System Journal **40(2)** (2001) 310-321
3. H.M.Berman, J.Westbrook, Z.Feng, G.Gilliland, T.N.Bhat, H.Weissig, I.N.Shindyalov, P.E.Bourne, The Protein Data Bank. Nucl. Ac. Res. **28** (2000) 235–242
4. T. L. Blundell. Structure-based drug design. Nature **384** (1996) 23–26
5. C. Branden and J. Tooze, *Introduction to Protein Structure.* Garland, 1999
6. J. M. Bujnicki, A. Elofsson, D. Fischer and L. Rychlewski. LiveBench-1: Continuous benchmarking of protein structure prediction servers. Prot. Sc. **10** (2001) 352–361

7. D. F. Burke, C. M. Deane, and T. L. Blundell. Browsing the SLoop database of structurally classified loops connecting elements of protein secondary structure. Bioinformatics **16(6)** (2000) 513–519
8. T. E. Creighton, *Proteins: Structures and Molecular Properties*. Freeman, New York, 1993
9. S. Cristobal, A. Zemla, D. Fischer, L. Rychlewski and Arne Elofsson. How can the accuracy of protein models be measured? (2000) submitted
10. J. Devillers (ed), *Genetic Algorithms in Molecular Modeling*, Academic Press, London, 1996
11. L. E. Donate, S. D. Rufino, L. H. Canard, and T. L. Blundell. Conformational analysis and clustering of short and medium size loops connecting regular secondary structures: a database for modeling and prediction. Protein Sci. **5(12)** (1996) 2600–2616
12. D. Fischer et al.. CAFASP-1: Critical Assessment of Fully Automated Structure Prediction Methods. Proteins **Suppl 3** (1999) 209–217
13. http://www.cs.bgu.ac.il/dlfisher/CAFASP2
14. A. Godzik. The structural alignment between two proteins: Is there a unique answer?. Prot. Sc. **5** (1996) 1325–1338
15. A. Godzik, J. Skolnick and A. Kolinski, A topology fingerprint approach to inverse protein folding problem. J. Mol. Biol. **227** (1992) 227–238
16. D. E. Goldberg. *Genetic Algorithms in Search, Optimization and Machine Learning*, Addison-Wesley, 1989
17. D. Goldman. *PhD Thesis*, U.C. Berkeley, 2000
18. D. Goldman, S. Istrail and C. Papadimitriou, Algorithmic Aspects of Protein Structure Similarity. *Proc. of the 40th IEEE Symposium on Foundations of Computer Science*, 512–522, 1999.
19. J. G. Gough, C. Chothia C., K. Karplus, C. Barrett and R. Hughey. Optimal Hidden Markov Models for all sequences of known structure. *Currents in Computational Molecular Biology* (Miyano S., Shamir R., Toshihisa T. eds.), Univ. Acad. Press Inc. Tokyo, 2000.
20. M. Grötschel, L. Lovász and A. Schrijver. The Ellipsoid Method and its Consequences in Combinatorial Optimization. Combinatorica **1** (1981) 169–197
21. T. F. Havel, G. M. Crippen and I. D. Kuntz, Biopolymers **18** (1979), 73
22. T. F. Havel, I. D. Kuntz and G. M. Crippen. The theory and practice of distance geometry. Bull. Math. Biol. **45** (1983) 665–720
23. R. B. Hayward, C. Hoang and F. Maffray. Optimizing Weakly Triangulated Graphs. Graphs and Combinatorics **5** (1987) 339–349
24. J. H. Holland. *Adaptation in Natural and Artificial Systems*. MIT Press, 1992
25. L. Holm and C. Sander. Protein Structure Comparison by Alignment of Distance Matrices. J. Mol. Biol. **233** (1993) 123–138
26. L. Holm and C. Sander. Mapping the protein universe. Science **273** (1996) 595–602
27. M. Khimasia and P. Coveney, Protein Structure prediction as a hard optimization problem: the genetic algorithm approach, 1997
28. Y. Lamdan, J. T. Schwartz and H. J. Wolfson. Object Recognition by Affine Invariant Matching. *Proc. of the IEEE Conference on Computer Vision and Pattern Recognition*, 335–344, 1988
29. G. Lancia, R. Carr, B. Walenz and S. Istrail, 101 Optimal PDB Structure Alignments: A Branch-and-Cut Algorithm for the Maximum Contact Map Overlap

Problem. *Proc. of the 5th ACM REsearch in COMputational Biology*, 193–202, 2001

30. C. Lemmen and T. Lengauer, Computational methods for the structural alignment of molecules. Journal of Computer–Aided Molecular Design **14** (2000) 215–232

31. A. M. Lesk. Computational Molecular Biology. In: Encyclopedia of Computer Science and Technology (A. Kent, J. Williams, C. M. Hall and R. Kent eds.) **31** (1994) 101–165

32. M. Levitt and M. Gerstein. A Unified Statistical Framework for Sequence Comparison and Structure Comparison. Proc. Natl. Acad. Sc. **95** (1998) 5913–5920

33. S. Lifson and C. Sander, Nature **282** (1979), 109

34. G. M. Maggiora, B. Mao, K. C. Chou and S. L. Narasimhan. Theoretical and Empirical Approaches to Protein–Structure Prediction and Analysis. In: Methods of Biochemical Analysis **35** (1991) 1–60

35. A. Marchler–Bauer and S. H. Bryant. Comparison of Prediction Quality in the Three CASPS. Proteins **Suppl 3** (1999) 218–225

36. McLaughlan, Acta Crystallogr. (1979)

37. L. Mirny and E. Domany. Protein fold recognition and dynamics in the space of contact maps. Proteins **26** (1996) 391–410

38. K. Mizuguchi, C. M. Deane, T. L. Blundell, M. S. Johnson and J. P. Overington. JOY: protein sequence-structure representation and analysis. Bioinformatics 14 (1998) 617-623

39. K. Mizuguchi, C. M. Deane, T. L. Blundell, and J. P. Overington. HOMSTRAD: a database of protein structure alignments for homologous families. Protein Sci. **7(11)** (1998) 2469–2471

40. J. Moult, T. Hubbard, S. Bryant, K. Fidelis, J. Pedersen and Predictors. Critical Assessment of Methods of Proteins Structure Prediction (CASP): Round II. Proteins **Suppl 1** (1997) dedicated issue

41. J. Moult, T. Hubbard, K. Fidelis, and J. Pedersen. Critical Assessment of Methods of Protein Structure Prediction (CASP): Round III. Proteins **Suppl 3** (1999) 2–6

42. A. G. Murzin, S. E. Brenner, T. Hubbard and C. Chothia. SCOP: a structural classification of proteins database for the investigation of sequences and structures. J. Mol. Biol. **247** (1995) 536–540

43. G. L. Nemhauser and L. Wolsey. *Integer and Combinatorial Optimization*, John Wiley and Sons, 1988

44. J. P. Overington, M.S. Johnson, A. Sali, and T.L. Blundell, Tertiary structural constraints on protein evolutionary diversity; Templates, key residues and structure prediction. Proc. Roy. Soc. Lond. **B 241** (1990) 132–145

45. K. Park, M. Vendruscolo and E. Domany, Toward an Energy Function for the Contact Map Representation of Proteins. PROTEINS: Structure, Function and Genetics **40** (2000) 237–248

46. A. Raghunathan. Algorithms for Weakly Triangulated Graphs, U.C. Berkeley Tech. Rep., CSD-89-503 (1989)

47. G. Rhodes, *Crystallography Made Crystal Clear*. Academic Press, 2nd Ed., 1999

48. R. B. Russell, G. J. Barton. Multiple protein sequence alignment from tertiary structure comparison. PROTEINS: Struct. Funct. Genet. **14** (1992) 309–323

49. A. Sali and T. L. Blundell. Definition of general topological equivalence in protein structures. A procedure involving comparison of properties and relationships

through simulated annealing and dynamic programming. J. Mol. Biol. **212(2)** (1990) 403-428

50. N. Siew, A. Elofsson, L. Rychlewski and D. Fischer, MaxSub: An Automated Measure for the Assessment of Protein Structure Prediction Quality. Bioinformatics **16** (2000) 776–785
51. A. P. Singh and D. L. Brutlag 1998. Protein Structure Alignment: A comparison of methods. Bioinformatics (2000), Submitted.
52. T. F. Smith and M. S. Waterman. Identification of common molecular subsequences. J. Mol. Biol. **147** (1981) 195–197
53. R. Sowdhamini, D. F. Burke, J. F. Huang, K. Mizuguchi, H. A. Nagarajaram, N. Srinivasan, R. E. Steward, and T. L. Blundell. CAMPASS: a database of structurally aligned protein superfamilies. Structure **6(9)** (1998) 1087–1094
54. M. J. Sutcliffe, I. Haneef, D. Carney and T. L. Blundell. Prot. Eng. **1** (1987) 377–384
55. S. Umeyama. Least–Squares Estimation of Transformation Parameters Between Two Point Patterns. *IEEE Transactions on Pattern Analysis and Machine Intelligence*, PAMI-13(4) (1991) 376–386
56. M. Vendruscolo, E. Kussell and E. Domany, Recovery of protein structure from contact maps. Fold. Des. **2** (1997) 295–306
57. M. Vendruscolo, R. Najmanovic and E. Domany, Protein Folding in Contact Map Space. Phys. Rev. Lett. **82(3)** (1999) 656–659
58. M. Vendruscolo, B. Subramanian, I. Kanter, E. Domany and J. Lebowitz, Statistical Properties of Contact Maps. Phys. Rev. E **59** (1999) 977–984
59. A. Zemla, C. Venclovas, J. Moult and K. Fidelis. Processing and Analysis of CASP3 Protein Structure Predictions. Proteins **Suppl 3** (1999) 22–29

Spatial Pattern Detection in Structural Bionformatics

Haim J. Wolfson

School of Computer Science, Tel Aviv University, Tel Aviv 69978, Israel
wolfson@post.tau.ac.il

1 Introduction

Geometric pattern detection and recognition appears as a major task in various fields such as Computer Vision ([27]), Biometrics ([35]), Medical Image Processing ([29]), Classification of Anatomical Form ([7]) and more. In the last decade a new major application of 3D geometric pattern discovery has emerged in the rapidly developing field of Bioinformatics ([60]), which is dealing with the development of algorithms for Molecular Biology applications.

In its early stages Bioinformatics dealt almost exclusively with pattern discovery in DNA and protein sequences ([58]), which resulted in the development of efficient algorithms for biological sequence and multiple sequence alignment ([28]). These algorithms have become an indispensable tool in the analysis and interpretation of novel genomic data supplied by the Human Genome project ([18]). One of the key applications of pattern discovery is in the elucidation of the function of novel proteins based on their global sequence similarity to proteins of known function. Another key application is the detection of short sequential patterns (motifs), which are responsible for a similar specific function in otherwise different proteins (e.g. the calcium binding EF motif described in chapter 2 of [8]).

Elucidation of protein function is one of the key tasks of Molecular Biology. Proteins are most versatile molecules, which are involved in major processes of living organisms, such as catalysis, metabolism, signaling, transport, regulation, detection and destruction of foreign invaders by the immune system, assembly of new proteins and more. To quote A.M. Lesk ([44]) "In the drama of life on a molecular scale, proteins are where the action is". Under natural conditions, which depend on temperature and the solvent, protein molecules possess (almost) unique stable three-dimensional structures, which are defined by their amino-acid sequence. This structural information is significantly more

C. Guerra, S. Istrail (Eds.): Protein Structure Analysis and Design, LNBI 2666, pp. 35-56, 2003.

informative for the deduction of protein function, than the sequence information alone. Since protein structure is more conserved (during evolution) than its sequence, we can detect proteins with less than 25% sequence similarity, yet with roughly similar structure and function. It should be noted, though, that sometimes functionally non related proteins might fold into similar stable structures. Among the 15,000 structures in the Protein Data Bank ([4], circa March 2002) there are only about 700 structurally different single chain protein folds. Since proteins function by association, the spatial arrangement of certain residues on the protein molecular surface is crucial for their function. The books by Lesk ([44]) and Branden & Tooze ([8]) give an excellent in-depth introduction into protein structure, while lively descriptions and illustrations can be found in the popular book of Goodsell ([26]).

Most of the protein structures known today have been detected by X-ray crystallography and NMR based techniques. These methods are time consuming and cannot be applied to all proteins due to physical constraints. The most challenging task is, naturally, to computationally detect the fold of novel proteins from basic principles. The crucial role of protein structure analysis in the elucidation of protein function has triggered the initiation of the Structural Genomics ([9]) project as a natural follow up of the Human Genome project. The aim of this project is to detect (mainly, by X-ray crystallography) the structures of representatives from each cluster of sequentially homologous proteins and, consequently, to model the structures of the other proteins in the respective clusters based on sequence similarity to the detected structural template. This project is expected to supply numerous novel protein structures for subsequent structural and functional analysis, which requires a computational infrastructure similar to the one that was developed for protein sequence analysis. This algorithmic infrastructure, which gained the name of *Structural Bioinformatics*, receives as input the 3D shapes of proteins and, thus, from the mathematical viewpoint belongs to the discipline of *Geometric Computing*, which includes Computational Geometry, Computer Vision, Robotics, Computer Graphics, Medical Imaging and more. In particular, the structural analog of the classical protein sequence alignment tool, which is protein structure alignment, is conceptually similar to the classic problems of Object Recognition in Computer Vision ([51]). Both are problems of spatial pattern detection.

In our discussion we shall distinguish between *pattern recognition* and *pattern detection* or *pattern discovery*. While the term *pattern recognition* will be used for the detection of an *a-priori* known pattern/template in a database of shapes, the term *pattern detection* will apply for the detection of *a-priori* unknown (structural) patterns, which appear in several shapes (or recurrently appear in a single shape) of the database.

In Structural Bioinformatics spatial pattern discovery dominates the task of protein structural comparison and the task of the prediction of protein-protein, protein-DNA or protein-drug interaction (docking). In the general case the individual molecular structures also posses internal degrees of freedom, so we are faced with pattern detection in *flexible* spatial structures. In this paper we shall shortly outline the major computational issues one is facing in the above mentioned tasks and survey some of the work done at the Tel Aviv University Structural Bioinformatics group ([72]) on these topics. We shall first discuss the shape representation issue of protein structures and what are the relevant shape representations for the different computational tasks, then we shall outline some rigid and flexible protein structural alignment algorithms, and discuss the issues of rigid bound versus unbound and flexible docking.

2 Protein Shape Representation

Different protein shape representations are used in biological pattern discovery. The choice of the proper representation depends (first of all) on the biological task at hand and (to a lesser extent) on the algorithmic methodology.

Like any molecule a protein can be viewed as a set of its individual atoms, where each atom type is modelled as a ball of a given radius. Since the lengths of these radii are restricted to a relatively small interval and there are also stringent restrictions on the minimal distance between the centers of these atomic balls, such molecular shapes can be very efficiently handled by various Computational Geometry algorithms dealing with intersection and location queries ([30]). The full volumetric representation is important in docking studies, where one is fitting a pair of molecules, as if they were separate pieces of a 3D jigsaw puzzle. There one tries, first, to detect complementary spatial patterns on the surfaces of the shapes ([14]), and in a later stage has to verify that the proposed fit does not cause intersection of the interiors of the two volumes.

A protein can also be viewed as a folded 3-D curve of amino acids, the, so called, polypeptide chain ([44]). Each amino acid is built from a central C_α atom to which are attached a hydrogen atom, carboxyl and amino groups and a specific side chain residue, which differs from one amino acid to the other. Usually, the proteins function via exposed surface residues, while the internal residues supply the structural scaffold. Thus, if our aim is to detect similar spatial patterns in the protein folds, we can often reduce a protein representation to the set of its C_α atom centers, thus representing each amino acid by a single 3D point. If the order of the amino-acids on the polypeptide chain is relevant to the task at hand, we further reduce a protein to a 3D curve sampled at the centers of its C_α atoms. This is, essentially, a one-dimensional structure. Whether one should use the non-connected 3D point constellations or could consider a protein as a curve depends on the biological problem that we have to address. If one is looking for similar functional patterns of residues appearing on the protein surface, then the unconnected 3D representation is appropriate, while for evolutionary classification of protein folds the 3D curve representation might be a natural choice.

A much coarser representation of the polypeptide chain can be achieved by viewing it as an ensemble of secondary structure elements. Each element is a regular structure of consecutive amino acids, which are formed in the protein interior. The most common secondary structures are the α-helices which have a helical structure and β-sheets, consisting of β-strands ([44]). Both the α-helix axes and the β-strands can be modelled as roughly linear segments of 10-15 amino acids. Thus, one can reduce by an order of magnitude the size of the point set representation by representing a protein as a set of line segments. Here, as well, one might consider either ordered or unordered sets, depending on the problem at hand ([38, 1]). Of course, one could consider mixed sets of points and segments.

A major issue in protein shape representation is the modelling of flexibility, or internal degrees of freedom. While the native protein structure is assumed to be rigid, under different physical conditions or in association with other molecules, the protein might change its 3D shape. One has to consider both major changes, like domain movements ([25]), or minor frequently occurring changes in the shape and location of the side chains lining the outer molecular surface. This problem is especially acute in pattern discovery tasks associated with the prediction of protein-protein interaction (docking). The tasks then resemble articulated object recognition in computer vision ([71]) or matching of non-rigid human organs in medical image processing.

To summarize, depending on the biological task at hand, proteins might have very different shape representations, which directly affect both the size of the input and the algorithmic complexity of handling these shapes. Pattern detection in flexible shapes is obviously much more complex than in rigid ones. Viewing a protein as a 3D curve, which is in essence a one-dimensional entity, enables application of significantly more efficient algorithms than the handling of a protein as a disconnected set of 3D points, and reduction to secondary structures reduces the input size by an order of magnitude. Yet, for the detection of functional patterns on protein surfaces and for docking prediction, one needs to consider inherently 3D shape representations.

3 Protein Structural Alignment

3.1 Alignment of Rigid Structures

Assuming that proteins are rigid structures, the structural alignment of a pair of proteins is defined by a three dimensional rigid transformation (rotation and translation) which optimally superimposes their shapes. Naturally, the meaning of optimal shape superimposition depends on the protein shape representation. For comprehensive reviews on protein structural alignment algorithms see [20, 42]. As was mentioned in section 2 proteins may undergo conformational changes. These are generated by rotational movements around covalent bonds. Hinge and shear movements of protein domains have been observed as a combination of such rotations ([25]). Thus, a more complicated task is to discover structural similarity among molecules modulo internal degrees of freedom as well.

There are several applications where structural alignment plays a key role. Among these applications are the classification of the proteins universe by shape similarity ([47, 32, 14]), and the detection of three dimensional structural patterns, so called, **structural motifs**, which imply similar protein function. Structural alignment has also key applications to Computer-Assisted Drug Design, where one is trying to detect or design drugs which snugly fit some functional active site of a protein (receptor) molecular surface. Detection of structural similarity between the active site of a protein and a molecular surface patch of a novel protein implies that the new protein is a candidate receptor for the same drug. Likewise, if one detects other drugs having a certain structural similarity of functional groups to the original one, these drugs become candidate inhibitors for the same receptor.

Let us formulate the rigid structural alignment task as a particular instance of pattern detection, which is often called **partial matching**:

Given two rigid sets of features in the 3-dimensional space and the group of rigid 3D transformations, the *partial matching task* is to recover a transformation (or several transformations), which superimpose a large enough subset of the first structure onto the second one.

Note that in partial matching not all the features of either structure will have a matched counterpart. Moreover, one does not know in advance which of the features will have a counterpart. One should also note that in biological applications two features (e.g. atomic centers) are considered superimposed if the distance between them (in some metric) is less than a predefined threshold. Another practical observation is that we are not necessarily looking for the maximally matching subsets. Due to inaccuracies and fuzziness of correspondence the "correct solution" might not achieve the maximal score of a given algorithm. Moreover, even non-maximal matching substructures might imply similarity in biological function. Thus, one is looking for large matching subsets (and not only for the maximal one) each of which can represent a correct solution.

In order to tackle the partial matching task, one has to address two problems - feature correspondence and corresponding feature superposition. The correspondence problem is the difficult one. Once a correspondence hypothesis is established, there are well known efficient methods to find the rigid transformation which superimposes them with minimal least squares distance among the matching feature points (see, e.g. [36, 33, 59]).

We have tackled the partial matching problems in Structural Biology by methods which were originally developed for the object recognition task in Computer Vision and Robotics [40, 71]. A major goal of computer vision systems is to efficiently perform model-based object recognition in cluttered scenes [5, 12]. Specifically, given a database of familiar objects (models) and a newly observed scene, the task is to detect all the appearances of the models in the scene as well as the transformation between their pose in the model database and the scene. The objects appearing in the scene may be partially occluded and additional objects not appearing in the database may clutter the scene. Objects are usually represented by feature sets, such as interest points, line segments, surface patches etc.

One can notice that there is a striking conceptual similarity between the two tasks. Actually, the model-based object recognition task, as formulated above, is analogous to the structural alignment of a newly discovered molecular structure against a database of such structures. The analog of the model database is the molecular structural database, and the analog of the newly observed scene is the new molecular structure. Partial occlusion and additional scene clutter correspond to the fact that we are looking for previously undefined structural patterns. This analogy is even more direct, when in Computer Vision one is dealing with data acquired by a 3-D sensor, such as a range sensor. In essence, both in Computer Vision and in Structural Biology we are faced with the task of spatial pattern detection or partial matching . Guided by this conceptual similarity we have introduced Computer Vision based object recognition/matching techniques into Structural Bioinformatics [51].

Due to the need to compare databases of protein and DNA sequences, algorithms for character string matching have been extensively applied in Molecular Biology (see e.g. [58, 28]). Most of the methods are based on the dynamic programming approach, although hashing techniques have been applied as well to speed up processing [46]. Consequently, dynamic programming is deeply rooted in computational molecular biology and there have been significant attempts to tackle the three-dimensional structural matching task using this technique [63].

In order to apply the dynamic programming method, one has to exploit the order of the amino acid residues on the polypeptide chain. Thus, as was mentioned in section 2, from a purely geometric viewpoint the problem tackled is not that of 3D point set set partial matching, but of 3D curve partial matching. This is a significantly easier problem, since a curve in any dimensional space is just a **one dimensional** entity. The location of a point on a curve is fully defined by only one parameter, which is the arc-length ([17]). We shall see below that the sequence order constraint allows us to tackle the flexible alignment task without prior knowledge of hinge positions in the proteins ([61]). However, sequence order dependent matching algorithms cannot detect geometric patterns which do not depend on such an order, especially, molecular surface motifs.

Let us very shortly outline the sequence order independent protein structural alignment technique ([51, 2]). To facilitate the exposition we explain our method for structural alignment of C_α atom sets. We consider rigid objects, which are constellations of points. Each such point is located at the center of a C_α atom and may have a label (e.g. a residue name, or a residue type).

The technique we present is geared towards the efficient comparison of a target molecule to a database of known molecules. The pairwise structural comparison problem is just a particular case, where the database consists of one model. To simplify the exposition we describe our method for the pairwise comparison of point sets.

The three major steps of our approach are: detection of seed matches; clustering of seed matches; extension and verification of the candidate matching hypotheses.

1) **Detection of seed matches.** This is the heart of the algorithm. The goal is to establish localized structural alignments, which we nickname *seed matches*. The rationale of the localized approach is twofold - first, there is a biological justification for a local search, since the amino acids of biologically relevant structural patterns (e.g. active sites) are predominantly in spatial proximity, and, second, localization reduces the complexity of the computationally intensive step of the algorithm.

   The Geometric Hashing Technique [40, 51, 2] is applied to generate these seed matches. The localization is achieved by limiting the Geometric Hashing only to points with pairwise distance below a certain predefined threshold.

   Seed matches which receive a relatively high Geometric Hashing score are retained for further processing. Such a seed match is represented by a list of matching pairs of atoms (match-list) and by a 3-D rigid transformation (rotation and translation). The number of matching pairs should be above a threshold (minimal match-list size). Each pair in the list specifies a correspondence between an atom from one structure and an atom from the other structure. The transformation represents the 3-D rotation and translation, which superimposes the atoms of the first structure onto the corresponding atoms of the second structure.

   Note that this step may produce several candidate seed matches which induce (almost) similar transformations, obtained from different localized subsets. Thus, the next step clusters all these matching atoms together.

2) **Clustering of seed matches.** In the second step seed matches representing almost identical transformations are detected, and their match-lists are merged. A rotation and translation of one set onto the other giving the minimal least squares deviation of the matching points is computed [36, 59]) .

3) **Extension and verification of match hypotheses.** The relatively reliable correspondence lists of the seed matches obtained after the clustering procedure are extended to contain additional matching pairs. First, the structures are superimposed according to the transformation determined by the seed match of the cluster. Then, pairs of atoms that are "close" enough after the superimposition become prospective additional matching pairs. Since proteins are usually quite dense in space, each atom from one structure may have several "close" neighbors in the other structure. To

choose the appropriate pairs a heuristic iterative matching algorithm is applied, which minimizes the sum of the distances between all the matched pairs.

Finally, the best extended matches are reported. The quality of the match is determined mainly by the number of the matching pairs of atoms and, in some extent, by the least squares distance between these matching atoms sets. Naturally, one would prefer to incorporate in the final ranking biological information, which could assist in the detection of the biologically meaningful alignments, even if they are not large in size.

The algorithm was applied to the detection of a structurally non redundant representation of the Protein Data Bank ([14]), as well as to the creation of a structurally non redundant dataset of protein-protein interfaces ([65]). The algorithm can be accessed via the WWW at ([72]).

3.2 Structural Alignment of Flexible Proteins

In the previous discussion we treated proteins as rigid shapes. However, proteins do possess internal rotational degrees of freedom. In particular, as classified by [25], large proteins, which are built of several domains, often have different conformations, which can be modelled as hinge or shear movements of respective domains. These movements are important for protein function, such as biomolecular recognition/docking. A flexible structural comparison algorithm would enable us to perform the same searches as a rigid one with the added strength of being able to detect partial matches, even if one of the matching substructures has undergone conformational changes. This ability is especially important in the search of drug molecules against a database, since the vast majority of molecules have many potential conformations.

Flexible Alignment with Pre-defined Hinge Locations

For the sake of clarity let us discuss the following problem: one molecule is rigid and the other is composed of two rigid parts (domains) which are joined by a hinge. A hinge in this case is a rotational joint with full 3-D rotational freedom. In practice, rotations are only around specific bonds, but we allow this more general 3-D rotation model since it implicitly allows approximation of several consecutive (or, nearby) bond rotations, as well. The fact that one molecule is rigid can be assumed without loss of generality, while the introduction of several hinges in the other molecule is a straightforward extension of the algorithm that we present. Our flexible partial matching method is based on the ideas that we have introduced for articulated (flexible) object recognition in Computer Vision [71, 3]. It exploits an associative memory indexing approach reminiscent of the one used in Geometric Hashing, except that this time we have the additional task of handling the internal flexibility. Of course, one can apply the rigid matching method to the individual rigid parts, and then check whether there is a pair of high scoring hypotheses, one for each part, which are spatially consistent with the single flexible object input. However, our aim is to handle the information obtained from all the molecule rigid parts in an integrated manner, so that the whole is more than just the sum of its parts. We achieve this goal by accumulating evidence on the position of the hinge location of the model molecule relative to the target molecule. Since the hinge belongs to both rigid parts and its location is not influenced by the internal rotation, both parts contribute evidence which is integrated in our algorithm.

Below we describe the algorithm in more detail. Assume that we have a database of flexible model molecules, each consisting of two rigid parts joined by a hinge at a known position. Given a target molecule, we want to find a large partial match of the target with some model molecule allowing both external rotation and translation and internal rotation at the hinge. Thus, we are trying to detect a flexible spatial pattern. The molecules are represented by their C_α backbone atoms.

We first preprocess the database of the flexible molecules, and then perform the pattern discovery (recognition) versus the target. For each database (model) molecule the following **preprocessing** is done:

(a) The molecule is represented by its C_α backbone atoms, as interest features.
(b) The (known) hinge location is chosen as the origin of a 3-D reference which is denoted as the 'hinge frame'. The orientation of this frame is set arbitrarily.
(c) Each ordered non-collinear triplet of atoms belonging to a single rigid part and satisfying both proximity and non-degeneracy constraints is defined as a 'feature group'. An unambiguous 3D reference frame is defined for the triangle. For example, given the non-collinear triplet of points p_0, p_1, p_2, one can set the origin of the reference frame at p_0. To set the reference set axes, let us define two vectors $v_1 = p_1 - p_0$; $v_2 = p_2 - p_0$. Define v_1 as the direction of the x-axis, the cross product $v_1 \otimes v_2$ as the direction of the z-axis, and the unit vector of the y-axis as the cross product of the unit vectors of the x and z axes.
(d) The (rotation and translation invariant) triple of triangle side lengths serves as an address to a (hash) table, where one stores the information on the model molecules, rigid part, feature group and, especially, on the transformation between the feature group reference frame and the hinge frame.

In the **pattern discovery/recognition** stage we repeat a similar procedure for the rigid target molecule exploiting the information accumulated in the associative memory:

(a) The target is represented by its C_α backbone atoms, as interest features.
(b) Non-collinear triplets satisfying the proximity and non-degeneracy constraints are tagged as feature groups. A 3D reference frame is defined for each feature group as above. The triangle side lengths invariants serve as an address to the table and the transformations stored there are applied to the feature group reference frame. This transforms the reference frame to a new frame, which is a candidate 'hinge frame'. Votes are accumulated for the location of candidate hinge frames.
(c) After completion of stage (b) we consider pairs *(model molecule, hinge location)*, which have received a high vote. (In an ideal situation, such a candidate solution should exhibit two clusters of internal rotations representing the groups of the different molecule parts.)
(d) High scoring solutions can be further evaluated by different physico-chemical criteria as well as criteria requiring participation of both molecule parts in the detected pattern etc. Finally, match-lists are derived and a best least-squares match for each rigid part is computed.

In this algorithm we have exploited the fact that both parts incorporate the same hinge by locating the global reference frame of the flexible model-molecule at the hinge. In such a way, both parts contribute votes to a reference frame at the same location, although at different orientations.

Although here, our algorithm is described for molecules with a single hinge, it is extended to multiple hinges by the following enhancement. Instead of having one 'hinge frame' one can define multiple 'hinge frames', each of them centered at a different hinge. In the preprocessing stage, for each feature group on a single part, one should encode the transformations from its frame to all the 'hinge frames', which are located on that part. Thus, e.g., on a part with two hinges, two transformations will be stored for each feature group, while on a part with one hinge only, one transformation will be stored. The recognition phase will remain as described above, except that each target feature group votes for as many frames as the number of different transformations stored in its table entry.

We have implemented variants of this technique for flexible protein and small molecule structural alignment ([69] and the *FlexMol* algorithm of [61]), and protein-ligand docking [55, 57].

Simultaneous Flexible Alignment and Hinge Location Detection

The previous exposition referred to the case where we have *a-priori* knowledge of the candidate hinge positions on one of the molecules, or enumerate these positions, if it is computationally feasible. To overcome this handicap, we have recently developed a novel algorithm, which automatically detects both the candidate hinge positions on one of the molecules and the largest aligned flexible spatial patterns. The algorithm, *FlexProt* ([61]) accepts as input two molecules, assuming that the first is rigid and the second is flexible (the situation is symmetric from the mathematical standpoint), and outputs lists of best alignments ranked according to the number of detected rigid parts, the size of the overall alignment, the size of the individual rigid parts, the RMSD of the alignment and some additional criteria. Thus, the algorithm overcomes the requirement of prior partition of one of the molecules to rigid parts. Yet, in order to accomplish this task efficiently, it exploits locally the amino acid sequence order and is not fully sequence order independent as the algorithm described above.

The goal of the *FlexProt* algorithm is to divide the two protein molecules into a minimal number of separate consecutive fragments of maximal size, such that the fragments which are matched will be almost congruent (ϵ-congruent). Two rigid fragments are ϵ-congruent, if they have the same number of C_α atoms and there exists a 3-D rotation and translation which superimposes the corresponding atoms with an RMSD less than some pre-defined threshold ϵ. The arrangement of the matching fragments should be consistent with their order on the protein chain. Flexible regions are located between the rigid matching fragments. A trivial way to achieve a flexible alignment of maximal size is to allow flexibility between each pair of neighboring amino acids (C_α atoms), thus aligning completely the two molecules, if they are of the same length. Thus, to avoid such absurd "optimal" solutions, our goal is to minimize the number of flexible regions or rigid fragments. Clearly, the two goals of maximal matching size and minimal number of flexible regions are conflicting and some compromise heuristic should be applied.

The input to the algorithm are two protein molecules M_1 and M_2, each being represented by the sequence of its C_α atom coordinates. Assume that molecule M_1 has undergone hinge bending movements at several locations along its backbone. Further assume that between the flexible joint regions there are fragments without a significant structural change. The resulting hinge-bent molecule is denoted M_2. Under our assumptions there exists a set of rigid fragments of M_2 that are ϵ-congruent to the corresponding set of fragments of M_1. The model presented applies not only to different conformations of a given molecule, but to the general case of flexible motif detection between two molecules with different sequences.

The algorithm has the following major steps:

1. Detect all the large enough ϵ-congruent rigid fragment pairs, one from each molecule. This is done by aligning each atom pair (one atom from each molecule) and extending the alignment to the left and to the right, along the backbone chain, until the RMSD of the superposition of the two fragments gets larger than ϵ. The procedure is efficient, since (based on the calculations of the previous steps) one can compute the RMSD of each extension in $O(1)$ time ([36, 59]). Thus, the procedure is linear in the size of the matched C_α atom pairs. As a by-product we also get the rotation and translation, which best aligns the pair of fragments. At the end of this step, we have recovered all the possible ϵ-congruent rigid fragment pairs together with the transformations, which superimpose them.

2. Construct long enough flexible alignments by concatenating matching rigid fragment pairs in a way that is consistent with the amino acid sequence order, and does not create too large gaps between consecutive fragments. This is done by a graph-theoretic method reminiscent of the one applied in the FASTA algorithm for sequence alignment ([28]). A graph is built with the congruent fragment pairs as its vertices. A directed edge joins a pair of vertices, if the fragments they represent might be consecutive in the final alignment. Weights are assigned to the edges, rewarding long matching fragments and penalizing big gaps as well as large discrepancies in the size of the gaps for the pair of proteins. A virtual (source) vertex, which is connected with a zero weight edge to each original vertex in the graph, is added and a *Single-Source Shortest Paths* algorithm ([16]) is applied to the graph starting from the virtual vertex. This algorithm detects the shortest paths from the virtual vertex to each other vertex in the graph. The number of vertices in each such path represents the number of rigid fragments in a potential flexible alignment. For each such number, the candidate solutions are sorted according to the total size and the RMSD of the fragment alignment.

3. For the leading candidate solutions the rotation and translation for each fragment pair is recovered, and consecutive fragment pairs are joined into *epsilon*-congruent regions, if their transformations are close enough. This may happen, if between the fragments, one had insertions/deletions which did not affect the overall 3D structure of the local region. Thus, the discovered flexible pattern, appearing in both proteins is represented by a number of *epsilon*-congruent region pairs, each consisting of one or of several *epsilon*-congruent fragment pairs.

The theoretical complexity of the algorithm is bounded by $O(n^4)$ (where n is defined as the size of the larger molecule), however in practice its performance is much faster. We have done numerous experiments with *FlexProt*, which can be accessed via the WWW at [72]. The method proved to be robust and highly efficient comparing a pair of flexible structures of about 300 amino acids each in average time of 7 seconds on a standard desktop PC (400MHz, 256 Mb RAM). For further reference see [61].

4 Protein-Protein Docking

Protein-protein interactions play a major role in all biological processes. In addition, protein-drug docking is a major tool in computer assisted drug design. The binding affinity of the molecules is affected mainly by electrostatic, hydrophobic and van der Waals interactions. Since these non-covalent interactions are weak and act at short distances, in order to be effective the interacting molecules have to be very close to each other. As a result shape complementarity of the interacting molecules becomes a necessary condition for docking. Thus, the majority of the docking methods are searching, first, for complementary spatial patterns on the molecular surfaces of the interacting molecules ([43, 31]). The spatial pattern detection problem in docking is very challenging, since molecules usually undergo conformational changes upon association. This is aggravated by the fact that the residues on the molecular surface, which take part in docking, are more flexible than the residues in the protein interior.

When evaluating docking algorithms, one should distinguish between the, so called, 'bound' and 'unbound' cases. In the 'bound' case we are given the co-crystallized complex of two molecules. The complex is artificially separated by randomly rotating and translating one of the molecules. Now, the goal becomes to reconstruct the original complex. No conformational changes are involved in the 'bound' case. Success in bound docking examples is a natural pre-requisite for any docking algorithm, yet such a success does not ensure adequate performance in 'real life' cases, which are 'unbound'. In the 'unbound' situation we are given two molecules in their native conformations. The goal of the algorithm is to predict the 'correct' structure of the complex. In the few tens of cases, where we do have an independently resolved structure of the complex, one can verify the quality of the algorithm's prediction. Most of the docking algorithms encounter difficulties with the 'unbound' case ([31]).

As was mentioned above, significant geometric surface complementarity is usually a prerequisite for successful docking. We shall discuss mainly this geometric part of the docking methodology, where the task is to detect large enough patches on the molecular surfaces of the docked molecules, which are of complementary shape. Assuming that the molecules do not undergo large conformational changes, we are faced again with the 3D rigid inexact *partial matching* task.

Some of the algorithms approach this task by direct enumeration of the six dimensional rigid transformation space, to detect a translation and rotation, which best superimposes one molecule onto the surface of the other. Most of these methods ([70, 66, 67, 68, 24, 10, 11]) use brute force search of the 3 rotational parameters and the Fast Fourier Transform technique ([37]) for fast enumeration of the translations. Such algorithms are computationally expensive. Other algorithms define discrete interest features on the molecular surfaces and apply a partial shape matching algorithm on these features in a way similar to protein structural alignment algorithms. The pioneering docking algorithm in this direction was suggested by Kuntz et al. [39], where the problem was reduced to the detection of large enough cliques in the, so called, docking graph. We have applied the Geometric Hashing technique ([21]) for rigid docking as well as a variant of Geometric Hashing and Pose Clustering ([48, 49]), which proved to be relatively robust even for unbound docking ([50]). The flexible pattern detection method described in section 3.2 was applied to the docking of flexible molecules ([56, 57]). In protein-drug docking, one is often looking for complementary patterns of hydrogen bond donors versus hydrogen bond acceptors ([52, 53]).

Most of the geometric docking algorithms can be roughly divided into the following major steps :

1. **Molecular Surface Representation** - a popular representation is that of the solvent accessible surface as calculated by Connolly ([14, 13]). Sparser discrete *interest features*, such a points and associated normals ([15, 45, 49]) are extracted in this stage.

2. **Focusing on candidate binding (active) sites** - in order to significantly reduce the number of false positives and reduce computation time it is desirable to focus *a-priori* on the approximate areas of the molecular surface, where binding is likely to appear. Such candidate binding sites are usually detected by biological and shape criteria. An excellent example of 'biologically' defined binding regions are the complementarity defining regions(CDRs) in antibodies ([44]). An example of a shape criterion is the binding of drugs and small ligands in the large cavities of a receptor. There one might restrict the receptor surface to be explored to such cavities ([39]).

3. **Complementary spatial pattern detection** - this is the heart of the geometric docking algorithm and usually applies a partial shape matching algorithm on a set of discrete features, such as surface points with associated normals, which should align in roughly opposite directions. The output of this step is a set of candidate rigid transformations, which dock one molecule to the other.

4. **Geometric Complementarity Scoring and Ranking** - since molecules cannot penetrate into each other, candidate transformations from the previous step are discarded if they cause a significant penetration. Minor penetrations are allowed to reflect conformational changes of the molecular surface upon docking. In this step one can also calculate the size of the detected complementary surface (size of binding site) and score the hypotheses according to this size. (The fact that candidate solutions can be discarded due to penetration supplies a powerful false positives filter, which does not exist in the structural alignment algorithms described in section 3.)

5. **Biological Scoring and Re-ranking**- in this step one would like to accept the high enough scoring hypotheses of the previous step and re-rank them according to a free-energy function ([10]), which could discriminate between the biologically valid hypotheses and decoy complexes, which exhibit only geometric complementarity. To date this step seems to be the real bottleneck in the performance of unbound rigid docking algorithms ([31]).

We have recently developed an efficient rigid docking algorithm, which performs relatively well in the unbound docking task ([19]). The geometric part of the algorithm is based on methodology developed for Computer Vision applications and it extends the ideas presented in ([22, 50]). The algorithm partitions the molecular surface of the molecules into convex, concave and flat local patches of almost equal size. Patches with higher probability of belonging to the binding site are considered, and complementary configurations of pairs of interest points with associated normals are detected. Alignment of such pairs induce rigid transformations, which are subsequently tested for shape penetration and scored by geometric complementarity. The use of surface patches reduces the number of potential docking solutions, while still (in most tested cases) retaining the correct transformation. The algorithm can treat receptors and ligands of variable sizes. It succeeds in docking of large proteins (antibody with antigen) and small drug molecules. The running times of the algorithm are on the order of seconds for small drug molecules and minutes for large proteins. The algorithm was tested on most of the known benchmark complexes (see e.g. [11]) with satisfactory results. Figure 4 illustrates a challenging antibody-antigen docking example. We briefly sketch the outline of this algorithm, which is presented in [19].

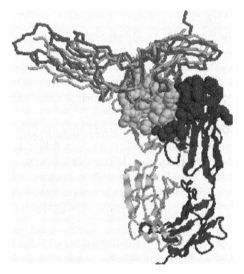

Fig. 1. Unbound docking of the Antibody Fab 5G9 with Tissue factor (PDB codes 1FGN,1BOY). The antibody is depicted by ribbons and the CDRs by atomic balls. The antigen is depicted as a bright backbone at the top of the figure. The dark (antigen) backbone represents the best solution obtained by our program (which was ranked 8'th by its score) superimposed on the complex with RMSD 2.27Å.

The algorithm follows the basic steps outlined above :

Molecular Surface Representation

We compute both a dense and sparse representation of the molecular surfaces for both each molecules. The dense surface representation is calculated using the MS program [14, 13], which outputs, concave, convex and saddle patches that are sampled with high density. In addition, the sparse surface representation of [45], which retains for each patch one point with its associated normal is calculated.

Segmentation of the Molecular Surface by Convexity

The input to this step is the sparse set of critical points. The goal is to divide the surface to patches of almost equal area. Each patch is of homogenous shape, namely, we segment the surface into convexities, concavities and flats. In order to decide the shape type in the vicinity of a single critical surface point we apply the shape function used in [15, 48]. A sphere is centered at each critical point and the shape function measures the ratio of the sphere volume, which is occupied by the molecule interior. The radius of the shape function sphere is selected according to the molecule size. We use 6Å for proteins and 3Å for small ligands. According to the histogram of the shape function values for the molecule, two cut-values are calculated, which split the critical points into three equal size sets of knobs (convex), flats and holes (concave). Using graph-theoretic techniques followed by a split-and-merge algorithm the molecular surface is divided into connected, almost equal size patches of concave, convex and flat points.

Detection of Potential Active/Binding Sites

A docking algorithm should successfully detect the binding sites of both molecules and their correct alignment. Knowledge of the binding site of at least one molecule greatly reduces the space of possible docking interactions, reducing both the run-time of the algorithm and the expected number of false positives. There are major differences in the interactions of different types of molecules. We develop filters for every type of interaction and focus only on the patches that were selected by the filter.

- *Protease-inhibitor interactions: Hot Spot Filtering.* A number of experimental and computational studies ([6, 34]) have shown that protein-protein interfaces have conserved polar and aromatic 'hot-spot' residues. We have used the results of these studies to select patches that have high probability of belonging to an active site. The other patches are discarded.
- *Antibody-Antigen interactions: Detection of CDRs.* It is well known ([44, 8]) that antibodies bind to antigens through their hypervariable (HV) regions, also called complementarity-determining regions (CDRs). The three heavy-chain and three light-chain CDR regions are located on the loops that connect the β strands of the variable domains. The CDRs are detected by aligning the sequence of the antibody to a consensus sequence of a library of antibodies. We restrict our docking algorithm to the patches of the CDR regions.

Matching of Complementary Patches

Given the patches of the pair of molecules we compute hypothetical docking transformations based on local geometric complementarity. We try to match convex patches with concave patches and flat patches with flat patches. We use two techniques - single patch versus single patch matching and pair of patches versus pair of patches matching. Single Patch Matching is used for docking of small ligands, like drugs or peptides, while Pairwise Patch Matching is used for protein-protein docking. Pairs of critical points with their associated normals are used as basic feature sets for matching by applying a combination of the Geometric Hashing ([40] and the Pose Clustering algorithm ([62]) techniques. Each such feature has a rotation and translation invariant **shape signature**, which includes the distance between the pair of points and the three angles that the normals and the distance segment define. Only features with almost similar shape signatures are aligned and the resulting transformation is stored. The method is similar to the one applied in [48]. The motivation in pairwise patch matching is that molecules interacting with big enough contact area must have more than one patch in their interface. Therefore matching two patches simultaneously will result in numerically more stable transformations. We consider only neighboring patches and choose the two points of the basic feature set from separate patches.

Pose Clustering.

Matching of local features may lead to multiple instances of similar pose. Therefore clustering is applied to reduce the number of potential solutions. We employ two clustering techniques. The first is clustering by the 6D transformation parameters, which is coarse, yet very fast, and thus applied first. The second is RMSD based clustering (similar to the one used in [54]), which is more exact but also much slower.

Steric Clash Detection and Ranking by Size of the Complementary Spatial Patterns

For each candidate transformation from the previous step, we superimpose the molecules according to the transformation. Now, we have to detect and discard those transformations, which cause unacceptable steric clashes (penetrations), and rank the rest of the solutions according to their shape complementarity. The detection of steric clashes and the scoring of shape complementarity requires to know the distance of the molecular surface points of one molecule from the molecular surface of the other. In general, points of one molecule, which penetrate deeply into the interior of the other discard the transformation, moderately penetrating points are penalized, and points, which are close to the molecular surface of the other molecule receive a positive score (binding site area). In order to accomplish this task efficiently, we compute in advance the distance transform of the first (stationary) molecule and calculate the score of each transformation using this distance transform. To further speed up the penetration detection and geometric scoring calculation, we construct a multi-resolution data structure for the transformed molecule. Only transformations that have not been rejected and received a large enough score at the lower resolutions, are re-scored at a higher resolution.

Refinement of the Candidate Transformations

The transformation from the matching step was computed based on aligning pairs points/ normals. Since the interface includes a much higher number of points, the transformation can be refined to improve geometric complementarity. Similar to the structural alignment algorithms described in section 3, for each candidate transformation a new extended match-list is compiled and the rotation and translation giving minimal RMSD for that match-list is calculated. This can be done in several iterations.

5 Summary

We have presented several applications of spatial pattern discovery algorithms to major tasks in Structural Bioinformatics, such as protein structural alignment and protein-protein docking. The exposition concentrated on algorithms developed by the interdisciplinary Structural Bioinformatics Group of Tel Aviv University and was not intended to review the enormous amount of work on this topic, which is done elsewhere.

Among the challenging tasks ahead of us is the development of efficient multiple structural alignment algorithms ([41]) and flexible multiple structural alignment algorithms. In docking, one should find better ways to handle surface side chain flexibility, and, especially, detect new ways for biological re-ranking of the geometrically ranked solutions. Our experience shows, that even in unbound rigid docking, we almost always detect a solution, which is very close to the 'correct' one, among the few hundred top ranked geome–tric solutions. This implies that one needs an efficient and robust biologically based score, which could successfully re-rank these top few hundred geome–trically derived hypotheses.

Successful docking algorithms might have direct implications on automated protein folding as well. If one could derive from the protein sequence partial structures of building blocks ([64]), then folding could be reduced to the docking of these building blocks subject to some proximity constraints. Work in this direction is underway.

Acknowledgments

I thank my long time colleague Ruth Nussinov and our joint graduate students, in particular, Dina Duhovny and Maxim Shatsky for valuable discussions and input. This research has been supported in part by the *Israel Science Foundation* administered by the *Israel Academy of Sciences* - Center of Excellence in "Geometric Computing", and by the Tel Aviv University Research Foundation.

References

1. V. Alesker, R. Nussinov, and H. J. Wolfson: Identification of non-topological motifs in protein structures. Protein Engineering. **9(12)** (1996) 1103–1119
2. O. Bachar, D. Fischer, R. Nussinov, and H. J. Wolfson: A computer vision based technique for 3-d sequence independent structural comparison. Protein Engineering. **6(3)** (1993) 279–288
3. A. Beinglass and H. J. Wolfson: Articulated object recognition, or, how to generalize the generalized Hough transform. In *Proc. of the 1991 IEEE Computer Society Conf. on Computer Vision and Pattern Recognition* (1991) 461–466
4. H. M. Berman, J. Westbrook, Z. Feng, G. Gilliland, T. N. Bhat, H. Weissig, I. N. Shindyalov, and P. E. Bourne: The protein data bank. Nucleic Acids Research. **28** (2000) 235–242
5. P. J. Besl and R. C. Jain: Three-dimensional object recognition. ACM Computing Surveys. **17(1)** (1985) 75–154
6. A. A. Bogan and K. S. Thorn: Anatomy of hot spots in protein interfaces. J. Mol. Biol. **280** (1998) 1–9
7. F. L. Bookstein: *Morphometric Tools for Landmark Data: Geometry and Biology.* Cambridge Univ. Press, Cambridge, UK 1991
8. C. Branden and G. Tooze: *Introduction to Protein Structure.* Garland Publishing Inc., New York 1991
9. S. K. Burley: An Overview of Structural Genomics. Nature Struct. Bio. (Structural Genomics Suppl.) (2000) 932–934
10. J. C. Camacho, D. W. Gatchell, S. R. Kimura, and S. Vajda: Scoring docked conformations generated by rigid body protein protein docking. PROTEINS: Structure, Function and Genetics. **40** (2000) 525–537
11. R. Chen and Z Weng: Docking unbound proteins using shape complementarity, desolvation, and electrostatics. ROTEINS: Structure, Function and Genetics. **47** (2002) 281–294
12. R. T. Chin and C. R. Dyer: Model-based recognition in robot vision. ACM Computing Surveys. **18(1)** (1986) 67–108
13. M.L. Connolly: Analytical molecular surface calculation. J. Appl. Cryst. **16** (1983) 548–558
14. M.L. Connolly: Solvent-accessible surfaces of proteins and nucleic acids. Science. **221** (1983) 709–713
15. M.L. Connolly: Shape complementarity at the hemoglobin $\alpha_1\beta_1$ subunit interface. Biopolymers. **25** (1986) 1229–1247

16. T. H. Cormen, C. E. Leiserson, and R. L. Rivest: *Introduction to Algorithms.* The MIT Press, 1990

17. M. P. DoCarmo: *Differential Geometry of Curves and Surfaces.* Prentice-Hall, Englewood Cliffs, N.J. 1992

18. U. S. DOE: Human genome project information. *http://www.ornl.gov/hgmis/publicat/publications.html*

19. D. Duhovny, R. Nussinov, and H.J. Wolfson: Efficient unbound docking of rigid molecules. *submitted*

20. I. Eidhammer, I. Jonassen, and W. Taylor: Structure comparison and structure patterns. J. Comp. Biol. **7(5)** (2001) 685–716

21. D. Fischer, S. L. Lin, H.J. Wolfson, and R. Nussinov: A geometry-based suite of molecular docking processes. J. Mol. Biol. **248** (1995) 459–477

22. D. Fischer, R. Norel, R. Nussinov, and H. J. Wolfson: 3-d docking of protein molecules. In *Fourth Symposium on Combinatorial Pattern Matching.* Lecture Notes in Computer Science **684** (1993) 20–34

23. D. Fischer, R. Tsai, R. Nussinov, and H.J. Wolfson: A 3-d sequence-independent representation of the protein databank. Protein Engineering. **8(10)** (1995) 981–997

24. H. A. Gabb, R. M. Jackson, and J. E. Sternberg: Modelling protein docking using shape complementarity, electrostatics, and biochemical information. J. Mol. Biol. **272** (1997) 106–120

25. M. Gerstein, A. M. Lesk, and C. Chothia: Structural mechanisms for domain movements in proteins. Biochemistry. **33(22)** (1994) 6739–6749

26. D. S. Goodsell: *Our Molecular Nature : the body's motors, machines, and messages.* Springer-Verlag, New York 1996

27. W. E. L. Grimson: *Object Recognition by Computer.* MIT Press, 1990

28. D. Gusfield: *Algorithms on Strings, Trees, and Sequences : Computer Science and Computational Biology.* Cambridge University Press, 1997

29. J. V. Hajnal, D. L. G. Hill, and D. J. Hawkes: *Medical Image Registration.* CRC Press, Boca Raton 2001

30. D. Halperin and M. H. Overmars: Spheres, molecules and hidden surface removal. Computational Geometry: Theory and Applications. **11(2)** (1998) 83–102

31. I. Halperin, B. Ma, H. J. Wolfson, and R. Nussinov: Principles of Docking: An Overview of Search Algorithms and A Guide to Scoring Functions. PROTEINS: Structure, Function and Genetics. **47** (2002) 409–443

32. L. Holm and C. Sander: Searching protein structure databases has come of age. PROTEINS: Structure, Function and Genetics. **19** (1994) 165–173

33. B. K. P. Horn: Closed-form solution of absolute orientation using unit quaternions. J. Opt. Soc. Amer. A. **4(4)** (1987) 629–642

34. Z. Hu, B. Ma, H.J Wolfson, and R. Nussinov: Conservation of polar residues as hot spots at protein–protein interfaces. PROTEINS: Structure, Function and Genetics. **39** (2000) 331–342

35. A.K. Jain, R. Bolle, and S. Pankanti(ed's): *Biometrics: Personal Identification in Networked Society.* Kluwer Acad. Pub., Boston 1999

36. W. Kabsch: A solution for the best rotation to relate to sets of vectors. Acta Crystallogr. **A32** (1976) 922–923

37. E. Katchalski-Katzir, I. Shariv, M. Eisenstein, A.A. Friesem, C. Aflalo, and I. A. Vakser: Molecular Surface Recognition: Determination of Geometric Fit

between Protein and their Ligands by Correlation Techniques. Proc. Natl. Acad. Sci. USA. **89** (1992) 2195–2199

38. I. Koch, T. Lengauer, and E. Wanke: An algorithm for finding maximal common subtopologies in a set of protein structures. J. Comp. Biol. **3** (1996) 289–306

39. I.D. Kuntz, J.M. Blaney, S.J. Oatley, R. Langridge, and T.E. Ferrin: A geometric approach to macromolecule-ligand interactions. J. Mol. Biol. **161** (1982) 269–288

40. Y. Lamdan and H. J. Wolfson: Geometric Hashing: A general and efficient model-based recognition scheme. In *Proceedings of the IEEE Int. Conf. on Computer Vision.* (1998) 238–249

41. N. Leibowitz, Z.Y. Fligelman, R. Nussinov, and H. J. Wolfson: Multiple structural alignment and core detection by geometric hashing. In *Int. Conf. on Intell. Systems for Mol. Bio.* (1999) 169–177

42. C. Lemmen and T. Lengauer: Computational methods for the structural alignment of molecules. J. Computer-Aided Molecular Design. **14(3)** (2000) 215–232

43. T. Lengauer and M. Rarey: Computational methods for biomolecular docking. Current Opinion in Structural Biology. **6** (1996) 402–406

44. A. M. Lesk: *Introduction to Protein Architecture : the structural biology of proteins.* Oxford Univ. Press, 2001

45. S. L. Lin, R. Nussinov, D. Fischer, and H. J. Wolfson: Molecular surface representation by sparse critical points. PROTEINS: Structure, Function and Genetics. **18** (1994) 94–101

46. D. J. Lipman and W. R. Pearson: Rapid and Sensitive protein Similarity Searches. Science. **227** (1985) 1435–1441

47. A. G. Murzin, S. E. Brenner, T. Hubbard, and C. Chothia: SCOP: a structural classification of proteins database for the investigation of sequences and structures. J. Mol. Biol. **247** (1995) 536–540

48. R. Norel, S. L. Lin, H. J. Wolfson, and R. Nussinov: Shape complementarity at protein-protein interfaces. Biopolymers. **34** (1994) 933–940

49. R. Norel, S. L. Lin, H. J. Wolfson, and R. Nussinov: Molecular surface complementarity at protein-protein interfaces: The critical role played by surface normals at well placed, sparse points in docking. J. Mol. Biol. **252** (1995) 263–273

50. R. Norel, D. Petrey, H. J. Wolfson, and R. Nussinov: Examination of shape complementarity in docking of unbound proteins. PROTEINS: Structure, Function and Genetics. **35** (1999) 403–419

51. R. Nussinov and H.J. Wolfson: Efficient detection of three-dimensional motifs in biological macromolecules by computer vision techniques. Proc. Natl. Acad. Sci. USA. **88** (1991) 10495–10499

52. M. Rarey, B. Kramer, and T. Lengauer: Time-efficient docking of flexible ligands into active sites of proteins. In *3'rd Int. Conf. on Intelligent Systems for Mol. Bio.* (1995) 300–308

53. M. Rarey, B. Kramer, T. Lengauer, and G. Klebe: A fast flexible docking method using incremental construction algorithm. J. Mol. Biol. **261** (1996) 470–489

54. M. Rarey, S. Wefing, and T. Lengauer: Placement of medium-sized molecular fragments into active sites of proteins. J. Computer-Aided Molecular Design. **10** (1996) 41–54

55. B. Sandak, R. Nussinov, and H.J. Wolfson: An automated robotics-based technique for biomolecular docking and matching allowing hinge-bending motion. Computer Applications in the Biosciences (CABIOS). **11** (1995) 87–99

56. B. Sandak, R. Nussinov, and H.J. Wolfson: Docking of conformationally flexible molecules. In *Seventh Symposium on Combinatorial Pattern Matching*. Lecture Notes in Computer Science **1075**. (1996) 271–287

57. B. Sandak, R. Nussinov, and H.J. Wolfson: A method for biomolecular structural recognition and docking allowing conformational flexibility. J. Comp. Biol. **5(4)** (1998) 631–654

58. D. Sankoff and J. B. Kruskal: *Time Warps, String Edits and Macromolecules: The Theory and Practice of Sequence Comparison.* Addison-Wesley, 1983

59. J. T. Schwartz and M. Sharir: Identification of Partially Obscured Objects in Two Dimensions by Matching of Noisy 'Characteristic Curves'. The Int. J. of Robotics Research. **6(2)** (1987) 29–44

60. J.C. Setubal and J. Meidanis: *Introduction to Computational Molecular Biology.* PWS, Boston, 1997

61. M. Shatsky, Z. Y. Fligelman, R. Nussinov, and H. J. Wolfson: Alignment of flexible protein structures. In *Int. Conf. on Intell. Systems for Mol. Bio.*, (2000) 329–343

62. G. Stockman: Object recognition and localization via pose clustering. J. of Computer Vision, Graphics, and Image Processing. **40(3)** (1987) 361–387

63. W. R. Taylor and C. A. Orengo: Protein structure alignment. J. Mol. Biol. **208** (1989) 1–22

64. C. J. Tsai, B. Ma, Y.Y. Sham, S. Kumar, H. J. Wolfson, and R. Nussinov: A hierarchical building-block-based computational scheme for protein structure prediction. IBM J. of Research and Development. **43(3-4)** (2001) 513–523

65. C.J. Tsai, S.L. Lin, H.J. Wolfson, and R. Nussinov: A dataset of protein-protein interfaces generated with a sequence order independent comparison technique. J. Mol. Biol. **260** **(4)** (1996) 604–620

66. I.A. Vakser: Protein docking for low resolution structures. Protein Engineering. **8** (1995) 371–377

67. I.A. Vakser: Main chain complementarity in protein recognition. Protein Engineering. **9** (1996) 741–744

68. I.A. Vakser, O.G. Matar, and C.F. Lam: A systematic study of low resolution recognition in protein–protein complexes. Proc. Natl. Acad. Sci. USA. **96** (1999) 8477–8482

69. G. Verbitsky, R. Nussinov, and H. J. Wolfson: Flexible structural comparison allowing hinge bending, swiveling motions. PROTEINS: Structure, Function and Genetics. **34** (1998) 232–254

70. P.H. Walls and J.E. Sternberg: New algorithms to model protein-protein recognition based on surface complementarity; applications to antibody-antigen docking. J. Mol. Biol. **228** (1992) 227–297

71. H. J. Wolfson: Generalizing the generalized Hough transform. Pattern Recognition Letters. **12(9)** (1991) 565–573

72. H. J. Wolfson and R. Nussinov: Structural Bioinformatics group. *http://bioinfo3d.cs.tau.c.il*

Geometric Methods for Protein Structure Comparison

Carlo Ferrari and Concettina Guerra

Department of Information Engineering, University of Padova,
Via Gradenigo 6a, 35131 Padova, Italy
tel: 039-049-8277821 fax: 039-049-8277826
{*carlo,guerra*}*@dei.unipd.it*

1 Introduction

Protein structural comparison is an important operation in molecular biology and bionformatics. It plays a central role in protein analysis and design. As proteins fold in three dimensional space, assuming a variety of shapes, a careful characterization of their geometry is needed to study their function which is known to be related to the shape. Moreover, the comparison of protein structures is essential to infer evolutionary information.

The problem of comparing three-dimensional structures has been widely studied in other disciplines such as computer vision and image processing, robotics, astronomy and some core methods have migrated from these disciplines to bionformatics.

There are many instances of the protein comparison problem that have been addressed; they include: 1) protein pairwise comparison, 2) protein classification, to organize all known structures in a biologically relevant groups, 3) searching for common folding patterns and three-dimensional motifs, 4) studying of protein interaction to identify binding sites for drug design.

From the application point of view, it is important to mention how the growth of the Protein Data Bank (PDB) asks for effective automatic procedures for classification and search of the database elements. Currently the PDB contains more than 17,000 structures and this number is rapidly growing.

The protein comparison may involve different levels of representations of the three dimensional protein structures, from the atomic level to the level of secondary structures. Most methods presented in the literature deal with a protein representation in terms of atomic coordinates

and therefore with a matching problem that uses as basic elements sets of points (atoms).

C. Guerra, S. Istrail (Eds.): Protein Structure Analysis and Design, LNBI 2666, pp. 57-82, 2003.
© Springer-Verlag Berlin Heidelberg 2003

Other approaches are based on secondary structures, i.e. α helices and β strands, that play an important role in the functional behavior of a protein. The secondary structure elements are represented as vectors in 3D space.

An alignment of α helices and β strands may be used for fast retrieval of folds or motifs from the PDB. On the other hand, the comparison of secondary structures can be used as the first step in a two step comparison procedure that first identifies possible candidate solutions in a fast way and then refines the solutions taking into consideration the atomic descriptions of proteins. Another advantage of a structural comparison of secondary structures is that it allows to study the folding process by tracing the evolution of the fold from the molden state.

Most of existing approaches allow to detect global similarity between entire proteins as well as local similarity ([1], [20], [21], [23], [26], [27], [41], [46], [51]).

The integration of strategies operating at different levels of representations appears very promising to achieve robustness and efficiency. Extensive surveys on the subject of protein comparison exist enphasizing different aspects of the general problem [6], [37], [50].

In this paper, we review some of the theoretical results on the computational complexity of the algorithms designed to obtain optimal solutions to the problem of matching sets of points using specific metrics. From a theoretical point of view, the problem has been extensively studied in the area of computational geometry, where it is often formulated as the problem of finding correspondences between sets of geometric features (for instance, points or segments). From these studies it appears that, in most practical cases, exact algorithms are too time consuming to be useful. Thus, approximate algorithms are considered that are computationally practical and at the same time are guaranteed to produce solutions that are within a certain bound from optimal.

Furthermore, we discuss methods for the estimation of rigid transformations under different metrics such as the Root Mean Square Deviation (RMSD) and the Hausdorff distance. Geometric indexing techniques prove their effectiveness in searching large protein databases and they are presented in details. Finally graph-theoretic protein modeling is reviewed as it is useful in designing algorithms for substructure identification and comparison.

Throughout the paper, we will use pure geometric information, ignoring other properties associated with atoms. Chemical properties, such as hydrophobicity, charge, etc. may be important in protein comparison and often they can be easily incorporated in a matching procedure. The use of such properties may help reduce the computation time by allowing pruning of the possible associations at early stages of the processing. However, we will not be consider these other properties in this chapter. Applications of protein comparison are an important subject; since they discussed in the two previous chapters of this volume, they are not considered here.

2 Protein Description

A protein is a sequence of aminoacids linked by peptide bonds. An aminoacid consists of a carbon atom C_α to which are attached a hydrogen atom, an amino group, and a carboxyl group. The 20 aminoacids differ in the side chain or *residue* attached to the C_α atom. The peptide bonds between the aminoacids in the chain join the carboxyl group of one aminoacid with the amino group of the next eliminating water in the linking process. The sequence of aminoacids is generally referred to as the *primary structure* of a protein. Its length varies from a few tens to few thousands aminoacids. A different level of protein representation, known as *secondary structure*, describes a protein in terms of recurrent regular substructures, such as the α *helices* and the β *strands*. The *tertiary structure* is the packing of the structural elements into the 3D shape. The protein may contain several chains forming its *quaternary structure*. For a survey of the protein architecture see [5], [38].

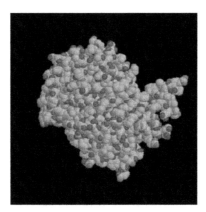

Fig. 1. The volumetric representation of protein 1rpa

The volumetric representation of a protein is displayed in figure 1 where all atoms are shown as balls; the secondary structure elements of the same protein are displayed as ribbons in figure 2.

Arrangements of the secondary structures α helices and β strands are the basis for the protein structural classification of SCOP [44]. In the SCOP classification hierarchy, the fold level corresponds to the last level of the hierarchy, the other two being family and superfamily. Proteins sharing a fold have the same major secondary structures but do not necessarily have a common evolutionary relationship, unlike proteins clustered into families and superfamilies. The similarity in the arrangements of secondary structures in a fold may be due to the physical and chemical properties of the packing of the proteins.

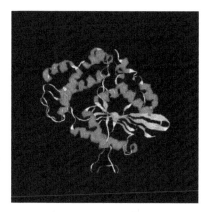

Fig. 2. The secondary structure representation of protein 1rpa

Fig. 3. The ribbon representation of protein 1bxa showing a β sandwich

Fig. 4. The ribbon representation of protein 3por forming a β barrel

Common structural arrangements of secondary structures or *motifs* have been identified within the folds and they include, among others, the *β-sandwich* and the *β-barrel*, as seen in figure 3 and 4, respectively.

Approaches to protein comparison use different protein structural descriptions. A complete structural description is given by the 3D coordinates (x, y, z) of the individual atoms of a protein. Often only the C_α atoms of the aminoacids, that form the so-called *backbone* of a protein, are considered for comparison.

A more compact description is in terms of the linear vectors associated to the structural elements helices and β strands. While most of the comparison approaches are based on the atomic description of a protein, the secondary structure description may provide a fast method to retrieve substructures or motifs from large protein databases. Furthermore, it is often used as a first step when searching in a database for the most similar protein with respect to a target protein. In fact, hypotheses of similarity for the target protein are generated in a fast and efficient way based on secondary structures only; such hypotheses are further verified by a more refined and costly process that is only applied to those hypothesized proteins. This two-step procedure may considerably speed up the protein comparison when large databases are involved.

For most proteins in the PDB, secondary structures are annotated by the original depositor who provides the starting and ending residue numbers of all secondary structures. However programs have been designed for the assignment of the secondary structures from the PDB files and for the analysis of the overall and residue-by-residue geometry of a protein [10], [34], [35].

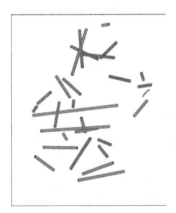

Fig. 5. The vectorial representation of protein kinase CK2

Several programs have also been developed to yield the vectorial representation of a protein [17], [41]. Singular-value decomposition (SVD) is a standard routine [4], [19] to find the axes of α helices and the best fit segments for the β strands. In this routine typically only C_α atoms are used. Other simpler methods derive the vector associated to a β strand either by directly connecting the starting and ending residues of the β strand assignments or by connecting two points that are computed as the average points of few of the extreme residues on both sides of the strand [51]. This second approach is less sensitive to curved or kinked structures. Figure 5 shows the vectorial representation of protein kinase CK2, where each segment is displayed as a cylinder of fixed radius.

3 Structural Comparison: Problem Formulation

The general matching problem can be informally defined as follows: Given two sets of geometric features, either points or line segments, determine the largest common subsets, i.e. the subsets of maximum size, that are geometrically similar. In the case of proteins, points correspond to atoms and segments are the axes of the secondary structures.

There are many variants of the matching problem that have been considered in many different contexts. We now give more formal definitions of the problem with varying degrees of computational complexity.

Problem 1. Consider two sets of geometric elements $A = \{a_1, a_2, \cdots, a_n\}$ and $B = \{b_1, b_2, \cdots, b_m\}$ in three-dimensional space and assume that they have the same cardinality, i.e. $n = m$, and that the element a_i corresponds to the element b_i. Find the transformation g between the two sets that minimizes a given distance metric D over all rigid body transformations T , i.e.

$$min_T D(T(A), B)$$

Problem 2. Consider two sets of geometric elements $A = \{a_1, a_2, \cdots, a_n\}$ and $B = \{b_1, b_2, \cdots, b_m\}$ in three-dimensional space. Find the transformation g between the two sets that minimizes a given distance metric D over all rigid body transformations T , i.e.

$$min_T D(T(A), B)$$

This problem differs from the previous one because no correspondence is known a priori between the elements of the two sets.

Problem 3. Given two sets $A = \{a_1, a_2, \cdots, a_n\}$ and $B = \{b_1, b_2, \cdots, b_m\}$ in three-dimensional space and a real number $\delta > 0$, find a maximum-cardinality set of pairs of elements, one element in A and the second one in B, such that the distance d between each pair of elements is at most δ.

In this problem we are interested in the largest subset of corresponding elements of A and B (generally, in practical applications the sets are required to be above a certain size).

Different metrics have been used in the literature to determine the structure similarity between geometric objects. The most common metric is the RMSD (Root Mean Square Deviation) defined for point sets as follows:

$$D(A, B) = RMSD(A, B) = (\textstyle\sum_{i=1,n} d(a_i, b_i)^2)^{1/2}$$

where d is the Euclidean distance between two points, and assuming that the sets have the same cardinality n, a_i corresponds to b_i.

The RMSD distance is useful when comparing very similar structures, as those produced during the christallographic analysis or NMR at different stages of the process. Its disadvantage becomes apparent in the presence of *outliers* when the proteins are not structurally close. The existence of even few outliers may significantly alter the value of the distance and therefore the determination of the optimal superposition of the two structures.

A second important definition between two point sets is based on the use of contact maps [33]. The chapter by G. Lancia and Sorin Istrail in this book deals extensively with contact maps and they are not further discussed here.

Another definition is the Hausdorff metric widely used in the area of computer vision and image processing, in astronomy and extensively studied in the field of computational geometry. The Hausdorff distance $H(A, B)$ between A and B is:

$$H(A, B) = \max(h(A, B), h(B, A)))$$

where $h(A, B)$ is the *one-way Hausdorff distance from A to B* given by:

$$h(A, B) = \max_{a_i \in A} \left(\min_{b_j \in B} d(a_i, b_j)\right)$$

In the following we discuss different approaches to solve the above three problems with different metrics. Problem 1 and its solutions are presented in section 5. Problem 2 with the Hausdorff distance as metric is considered in section 6.Problems 3 is reviewed in sections 8.

4 Representation of Rigid Transformations

A large number of methods have been proposed in the literature to compute the rigid body transformation between two sets of 3D points. They differ with respect to the transformation representation, and the minimization procedure. A survey by Sabata and Aggarwal [49] lists several representation of transformations and approaches to solve this problem using both closed form solutions and iterative solutions. The book [16] gives a clear description of many issues related to rigid transformations with enphasis on visualization aspects.

Here we describe several representations of rigid transformations while the next section is devoted to review methods to compute them.

A rigid motion of an object is a motion that preserves the distances between object points. The net movement of a rigid body from one configuration to another configuration (via a rigid motion) is called a *rigid displacement*. A rigid transformation applied to an object model represents a rigid displacement of the object itself. Rigid transformations form the theoretical infrastructure both for studying actual objects motions and for predicting possible (or hypothetical) motions. In the early 1800s Chasles and Poinsot proved that every rigid body displacement can be realized by a rotation about an axis combined with a translation parallel to that axis. This motion is what it is usually referred as a *screw motion*. Different motion representations are presently used (mainly in computer vision, computer graphics and robotics), that can be roughly classified as local or global representations [43].

Let us start by formally introducing the definition of *rigid body transformation*. A rigid body transformation is a mapping $g : \mathbb{R}^3 \to \mathbb{R}^3$, that must satisfy the properties:

$\| g(q) - g(p) \| = \| q - p \|$ for all points $p, q \in \mathbb{R}^3$
$g(v \times w) = g(v) \times g(w)$ for all vectors $v, w \in \mathbb{R}^3$

The former condition says that lengths are preserved and the latter condition says that internal reflection is not allowed. As a consequence of the above definitions, rigid body transformations also preserve the inner product, in particular, orthogonal vectors are transformed to orthogonal vectors. In general, a rigid body transformation takes right-handed orthonormal coordinate frames to right-handed orthonormal coordinate frames. It is important to point out that even if the distance between points and the cross product between vectors are fixed, particles in a rigid body can move related to each others, because they can rotate (but not translate) with respect to each other. Then the motion of a body can be described by the motion of any one point and the rotation of the body around this point. Hence a right-handed Cartesian coordinate frame can be attached to some point of the body and the motion of individual points can be traced from the motion of the body frame and the motion of the frame attachment point. Due to its importance we first consider pure rotational motion.

Pure rotational motion in $\mathbb{R}^3$ can usually be described by a proper 3x3 matrix, that can be defined by stacking next to each other, the coordinates of the principal axes of a coordinate frame B (the body frame) relative to a coordinate frame A (the inertial frame). Such a matrix is called a *rotational matrix*: its columns are mutually orthonormal and its determinant is +1. The set of all 3x3 matrices that satisfy these two conditions is denoted by $SO(3)$, where SO stands for *Special Orthogonal*. $SO(3)$ is a *group* under the matrix multiplications, with the identity matrix I as the identity element. $SO(3)$ is the *rotation group* of $\mathbb{R}^3$. A 3x3 rotation matrix can be seen as

$$\begin{bmatrix} r_{11} & r_{12} & r_{13} \\ r_{21} & r_{22} & r_{23} \\ r_{31} & r_{32} & r_{33} \end{bmatrix}$$

with nine different parameters, the orthonormality conditions (for the matrix columns) add three more constraints, while the sign of the cross product adds other three constraints. These six constraint equations, reduce the degree of freedom to three, that is, only three parameters are needed to completely represent a pure rotation in $\mathbb{R}^3$. The matrix coefficients r_{ij} can be expressed in term of these three parameters.

A rotation matrix $R \in SO(3)$ represents a rigid body transformation. In fact it can be proved that it preserves distances and orientations, that is:

- $\| Rq - Rp \| = \| q - p \|$ for all points $p, q \in \mathbb{R}^3$
- $R(v \times w) = Rv \times Rw$ for all $v, w \in \mathbb{R}^3$

Moreover a rotation matrix can be seen as an operator that takes the coordinates of a point (or vector) from a frame to another. Let p_b the coordinate of a point P with respect to the frame B, and R_{ab} the rotation matrix, the coordinates of P with respect to the frame A are given as:

$$p_a = R_{ab}p_b$$

The pure rotation operator is a linear operator (with the additional constraint that it is orthonormal). A sequence of two (or more) rotations will result in a single combined rotation and conversely a given rotation can be decomposed using two or more rotations. Rotation matrices can be combined to form new rotation matrices using matrix multiplication. If a frame C has orientation R_{bc} relative to frame B and B has orientation R_{ab} from frame A, then the orientation of C with respect to A is given by:

$$R_{ac} = R_{ab}R_{bc}$$

In particular, as we could expect, $R_{ab}R_{ba} = I$ and $R_{ba} = R_{ab}^{-1} = R_{ab}^{T}$.

An important result about rotations is the Euler Theorem that establishes that any rotation $R \in SO(3)$ is equivalent to a rotation about a given axis $w \in \mathbb{R}^3$ ($\| w \| = 1$), by an angle $\theta \in [0, 2\pi)$. In fact, it is possible to represent the motion of a single point p rotating about w at a constant unit velocity, with the following differential equation:

$$\dot{p}(t) = w \times p(t) = \hat{w}p(t)$$

Solving this equation gives the expression for a single rotation about w by θ, that is $R(w, \theta) = e^{\hat{w}\theta}$. The matrix $\hat{w}$ is defined as follows:

$$\begin{bmatrix} 0 & w_3 & -w_2 \\ -w_3 & 0 & w_1 \\ w_2 & -w_1 & 0 \end{bmatrix}$$

where $w^T = [w_1, w_2, w_3]$ and it has the property that $\hat{w}^T = -\hat{w}$. Such a matrix is a skew-symmetric matrix. If $\| w \| = 1$, $\hat{w}$ is a unit skew-symmetric matrix. It can be proved that the exponential $e^{\hat{w}\theta}$ can be rewritten in term of the skew-symmetric matrix, resulting in the so-called *Rodriguez's formula*:

$$e^{\hat{\omega}\theta} = I + \hat{\omega}\sin\theta + \hat{\omega}^2(1 - \cos\theta)$$

This method represents rotations through the *equivalent axis representation* . This is perhaps the most intuitive way of representing rotations. However, it has some disadvantages. The representation is not unique: in fact choosing $\omega' = -\omega$ and $\theta' = 2\pi - \theta$ gives the same rotation as ω and θ, being the exponential map many-to-one. Moreover, singularities occur when $\theta = 0$ and the exponential equals I, in such a case ω cannot be determined. These singularities create problems when computing the rotations. Finally the transformations resulting from the composition of multiple rotations cannot be easily computed.

Besides the canonical coordinates, there exist different coordinate systems for representing the rotation group, mainly used in robotics systems. The following method of describing the orientation of the coordinate frame B relative to A uses the *Euler angles*. At the beginning frames A and B are coincident. First, the frame B is rotated about the z-axis by an angle α, then it is rotated about the (new) y-axis by an angle β and finally B is rotated about the z-axis by an angle γ. The triple of angles (α, β, γ) represents the overall rotation, and the angles α, β, γ are called the *ZYZ Euler angles*.

These three rotations occurs at principal axes and the global rotation matrix can be computed from the three rotation matrices related to the three elementary rotations, that is, giving the specific values for α, β, γ it is easy to compute both R_{ab} and R_{ba}. The converse question of whether the map from SO(3) to α, β, γ is surjective is important. It can be proved that for any $R \in SO(3)$ it is possible to determine the Euler angles. This representation suffers from the problem of singularity at R $= I$.

In order to solve this problem new methods should be studied. Rotations in a 2D space can be represented by complex numbers on the unit circle. When moving to a 3D space, it is possible to generalize this idea, by introducing *quaternions* . Formally a quaternion Q is a 4-tuple of the form $< q_0, q_1, q_2, q_3 >$: where q_0 is the *scalar* component of Q and $\overrightarrow{q} = (q_1, q_2, q_3)$ is the *vector* component of Q. Hence, $Q = (q_0, \overrightarrow{q})$ with $q_0 \in \mathbb{R}$ and $\overrightarrow{q} \in \mathbb{R}^3$. The set of quaternions is a 4D vector space over the reals and it forms a group with respect to quaternion multiplication (denoted ".".). Quaternions multiplication is defined as follows:

$$Q \cdot P = (q_0 p_0 - \overrightarrow{q} \cdot \overrightarrow{p}, \; q_0\overrightarrow{p} + p_0\overrightarrow{q} + \overrightarrow{q} \times \overrightarrow{p})$$

The *unit* quaternions are the subset of all quaternions Q such that $\| Q \| = 1$, *where* $\| Q \|^2 = q_0^2 + q_1^2 + q_2^2 + q_3^2$

Each rotation matrix $R = e^{\hat{\omega}\theta}$ correspond to a unit quaternion defined as $Q = (\cos(\theta/2), \omega\sin(\theta/2))$. It can be proved that if Q_{ab} correspond to a rotation of frame A to B and Q_{bc} correspond to a rotation of frame B to C, then the rotation between frame A to C is given by the quaternion $Q_{ac} = Q_{ab} \cdot Q_{bc}$. An alternative representation of rotations, often used in computer vision, is the *unit quaternion*, Given a unit quaternion $Q = (q_0, \overrightarrow{q})$ the corresponding rotation is given by $\theta = 2\cos^{-1}(q_0)$ and $\omega = k\vec{q}$ with $k = 1/\sin(\theta/2)$, if $\theta \neq 0, \omega = 0$ otherwise.

The 3x3 rotation matrix R in terms of the unit quaternion is directly given by:

$$R = \begin{bmatrix} q_0{}^2q_1{}^2 - q_2{}^2 - q_3{}^2 & 0 & 0 \\ 0 & 1 & 0 \\ 0 & 0 & 1 \end{bmatrix}$$

Quaternions give a global parametrization of SO(3), at the cost of using four numbers instead of three to represent a rotation. Since the quaternions space has a group structure that directly corresponds to that of rotations, they provide an efficient representation without suffering from singularities.

Rigid body displacements usually are not limited to pure rotations (even if pure rotations represent an important subset), but they generally consist of rotations and translations. Pure translations have a very simple representation: given two co-oriented frames A and B, with p_{ab} the representation in A of the origin of B, for any point $q \in \mathbb{R}^3$, $q_a = p_{ab} + q_b$. Pure translations can be represented by 3D vectors.

A rigid body motion (a rigid body displacement) then can be represented by $p_{ab} \in \mathbb{R}^3$ and $R \in SO(3)$. The Cartesian product of $\mathbb{R}^3$ with SO(3), represents all the rigid body motions and it is denoted as SE(3) (that stands for *special Euclidean group*):

$$SE(3) = \{(p, R) : p \in \mathbb{R}^3, R \in SO(3)\}$$

Each element of SE(3) serves both as a specification of a rigid body placement (with respect to a fixed environment frame) and as a transformation taking the coordinates of a point from one frame to another. If a is a 3D point and a' is its corresponding point after a rigid transformation is applied, then the following relation holds:

$$a' = R_{ab}a + p_{ab}$$

The above can be expanded into:

$$\begin{bmatrix} a'_x \\ a'_y \\ a'_z \end{bmatrix} = \begin{bmatrix} r_{11} & r_{12} & r_{13} \\ r_{21} & r_{22} & r_{23} \\ r_{31} & r_{32} & r_{33} \end{bmatrix} + \begin{bmatrix} p_x \\ p_y \\ p_z \end{bmatrix}$$

The transformation of points can be usefully represented using the *homogeneous representation* . The homogeneous representation maps points and vectors in a 4D space, by adding a forth coordinate. The homogeneous coordinates of a point $q = (q_1, q_2, q_3)$ are $\bar{q} = (q_1, q_2, q_3, 1)$, while the homogeneous coordinates of a vector $v = (v_1, v_2, v_3)$ are $\bar{v} = (v_1, v_2, v_3, 0)$. Then a rigid transformation becomes:

$$\bar{a}' = \begin{bmatrix} a' \\ 1 \end{bmatrix} = \begin{bmatrix} R_{ab} & p_{ab} \\ 0 & 1 \end{bmatrix} \begin{bmatrix} a \\ 1 \end{bmatrix} = \bar{g}_{ab}\bar{a}$$

The 4×4 matrix '$\bar{g}_{ab}$ is called homogeneous representation of $g = (p_{ab}, R_{ab}) \in SE(3)$. Within homogeneous representation we obtain a linear representation of rigid body motions, and the standard matrix multiplication is the composition rule for rigid body motion.

Homogeneous coordinates are useful for representing *twists* . A twist is a couple $(v, \hat{\omega})$, with $v \in \mathbb{R}^3$ and $\hat{\omega}$ a skew-symmetric matrix. A twist can be written as:

$$\hat{\xi} = \begin{bmatrix} \hat{\omega} & v \\ 0 & 0 \end{bmatrix}$$

Of course $\hat{\xi} \in \mathbb{R}^{4x4}$, while the twist coordinates $\xi = (v, \omega) \in \mathbb{R}^6$. It can be proved that the exponential of a twist multiplied by a scalar, is an element of SE(3), that is, it represents a rigid transformation. Conversely it can be proved that every rigid transformation can be written as the exponential of some twist, that is, given a $g \in SE(3)$, there exists a twist $\hat{\xi}$ and $\theta \in \mathbb{R}$ such that $g = \exp(\hat{\xi}\theta)$. It is important to remind the reader that the exponential map is many-to-one and hence the choice of ω and θ may not be unique for solving the rotational component of the motion.

The concept of twist helps us to show that every motion is a *screw motion*. A screw motion is a motion which consists of a rotation about an axis in space by an angle of θ, followed by a translation along the same axis by an amount d. This motion is called a screw motion since it remind us to the actual motion of a screw that rotate and translates about the same axis. A screw consists of an axis l, a pitch h and a magnitude M, while a screw motion represents rotation by an amount $\theta = M$, about the axis l followed by a translation parallel to the axis l by an amount $h\theta$. The overall rigid displacement can be computed and it results as:

$$g = \begin{bmatrix} e^{\hat{\omega}\theta} & (I - e^{\hat{\omega}\theta})q + h\theta\omega \\ 0 & 1 \end{bmatrix}$$

If we choose $v = -\omega \times q + h\omega$, the twist coordinates $\xi = (v, \omega)$ generate the given screw motion. Going one step further it can be proved that it is possible to define a screw associated with every twist. Finally it is possible to conclude that every rigid body motion can be realized by a rotation about an axis combined with a translation parallel to that axis.

5 Determination of 3D Rigid Transformations

In this section we review solutions for problem 1, as formulated in section 3. Consider two sets of points $A = \{a_1, a_2, \cdots, a_n\}$ and $B = \{b_1, b_2, \cdots, b_m\}$ in 3D space with the same cardinality. Assume that point correspondences are known, i.e. that point a_i corresponds to the point b_i. The problem is to derive the rigid body transformation that optimally maps A into B. This problem is known as the *absolute orientation problem*. Points measurements are affected by some noise and errors, due both to the estimation of the point coordinates and to the determination of the point correspondences. Formally speaking this problem can be stated as a minimization problem, according to the least square error criterion.

In the literature there exist a large numbers of algorithms that compute 3-D rigid transformation between two sets of geometrical features. For our purposes, it is interesting to review some of the most popular closed-form solution using correspondent points [39]. The rotational and translational components are computed as solutions to a least square formulation of the problem. The proposed approaches differ mainly in the transformation representation. The first was developed by Arun, Huang and Blostein [4] and it is based on computing the singular value decomposition (SVD) of a matrix.

The problem has been formalized as a particular version of the well-studied *orthogonal Procrustes problem* , that can be stated as it follows:

$$\text{minimize } \| A - BR \| \text{ with respect to R, subject to } R^T R = I$$

where $A, B \in \mathbb{R}^{m \times 3}$ are given by the set of 3-D vectors a_i, and the set of 3-D vectors b_i, respectively, and $R \in \mathbb{R}^{3 \times 3}$, is orthogonal.

This problem is equivalent to the following one:

$$\text{maximize } trace(R^T B^T A) \text{ with respect to R, subject to } R^T R = I$$

In fact

$$(A - BR)(A - BR)^T = trace(A^T A) + trace(B^T B) - 2trace(R^T B^T A)$$

This problem can be approached using "The Singular Value Decomposition (SVD) Theorem", that can be stated as follows: **Theorem.** *If A is a real m-by-n matrix then there exist two orthogonal matrices $U \in \mathbb{R}^{m \times m}$ and $V \in \mathbb{R}^{n \times n}$ such that $U^T AV = diag(\sigma_1, \ldots, \sigma_p), p = min\{m, n\}$ where $\sigma_1 \geq \sigma_2 \geq \ldots \geq \sigma_p \geq 0$.* The problem of maximizing $trace(R^T B^T A)$ can be solved through the computation of the SVD of $B^T A$. In fact let Σ be the SVD of $B^T A$, that is

$$\Sigma = U^T (B^T A)V = diag(\sigma_1, \sigma_2, \sigma_3)$$

and we define a new orthogonal matrix $Z = V^T R^T U$. Then we obtain: $trace(R^T B^T A) = trace(R^T UU^T (B^T A)VV^T = trace(R^T U\Sigma V^T)$ then

$$trace(R^T U \Sigma V^T) = trace(V^T R^T U \Sigma V^T V = trace(Z\Sigma)$$

The previous expression can be rewritten as:

$$trace(R^T U \Sigma V^T) = trace(Z\Sigma) = \sum_{i=1}^{3} z_{ii}\sigma_i$$

as Z is orthogonal the best choice for R is obtained when Z becomes the identity matrix. Hence $R = UV^T$

This general method proves itself useful for computing 3-D rigid transformation estimation between two sets of corresponding points. The algorithm can be sketched as a two step algorithm. The first step computes the optimal rotation matrix R using the 3x3 correlation matrix $H = \sum_{i=1}^{N} a_i b_i^T$ through its singular value decomposition ($H = U diag(\sigma_1, \sigma_2, \sigma_3) V^T$) obtaining $R = VU^T$. The second step computes the optimal translation vector as $P = a - Rb$.

A similar approach computes the eigenvalues of a proper derived matrix (the *orthonormal matrices*) instead. It was proposed by Horn, Hilden and Negahdaripour [25]. With this method the correlation matrix H is firstly computed. However, rather than computing its SVD, a *polar decomposition* is used, such that $H = RS$, where $S = (HH^T)^{1/2}$. The optimal rotation is given by:

$$R = H^T \left(\frac{1}{\sqrt{\lambda_1}} u_1 u_1^T + \frac{1}{\sqrt{\lambda_2}} u_2 u_2^T + \frac{1}{\sqrt{\lambda_3}} u_3 u_3^T \right)$$

where $\{\lambda_i\}$ and $\{u_i\}$ are the eigenvalus and eigenvectors of the matrix HH^T.

Representing rotations using unit and dual quaternions gives two more techniques that have been proposed respectively by Horn [28] and by Walker, Shao and Voltz [56]. The former method asks to rewrite the minimization problem in the quaternion framework. A new 4x4 matrix can be constructed from the correlation matrix H as:

$$K = \begin{bmatrix} H_{00} + H_{11} + H_{22} & H_{12} - H_{21} & H_{20} - H_{02} & H_{01} - H_{10} \\ H_{12} - H_{21} & H_{00} - H_{11} - H_{22} & H_{01} + H_{10} & H_{20} + H_{02} \\ H_{20} - H_{02} & H_{01} + H_{10} & H_{11} - H_{00} - H_{22} & H_{12} + H_{21} \\ H_{01} - H_{10} & H_{20} + H_{02} & H_{12} + H_{21} & H_{22} - H_{11} - H_{00} \end{bmatrix}$$

The optimal rotation is the eigenvector related to the largest positive eigenvalue of K.

The latter method is the most significantly different of the four. It was designed ot minimize the equation;

$$\Sigma^2 = \sum_{i=1}^{L} \alpha_i \| n_1 i - R n_2 i \|^2 + \sum_{i=1}^{N} \beta_i \| a_i - R b_i - P \|^2$$

where $\{n_{1i}\}$ and $\{n_{2i}\}$ are two sets of corresponding unit normal vectors, and $\{\alpha_i\}$, $\{\beta\}$ are weighting factors reflecting data reliability. Dual quaternions for representing both rotation and translation are used and again the minimization problem can be rewritten in this new framework, resulting in new equations involving the parametrization of the dual quaternions. Again optimal values for R and P can be computed.

These four algorithms can be compared with respect to their accuracy, stability and efficiency [40]. Experimentations shows that no one algorithm is superior in all case. In fact difference in accuracy (on nondegenerate 3-D point sets) is almost insignificant. Stability is more discriminant instead. The SVD and the unit quaternion method are very similar and usually the most stable. In terms of efficiency, the orthonormal matrix looks quicker with small data sets, while the dual quaternions method is superior with larger data sets (according to proper computer memory configuration). In conclusion, the SVD should provide the best stability and accuracy, even if is not as efficient as dual quaternions with large data set. Otherwise the unit quaternions can be chosen on smaller data set for slightly better speed performance.

6 Geometric Pattern Matching

In this section we focus on the Hausdorff distance, defined in section 3, and review both theoretical and practical approaches to compute it.

Much work has been done on the computation of the Hausdorff distance. In the area of computational geometry exact algorithms have been studied for the problem of deciding whether there exist a transformation that maps one set of points into another set within a given distance. Fundamental robustness issues are discussed in [3]. Chew et al. [10] have considered the problem of matching point-sets in a d-dimensional space using the Hausdorff distance under translation only. For the case $d = 3$, they provide exact solutions in $O(n^3 \log^2 n)$ time. Extensions [11] to the more general case of Euclidean motion and of sets of segments have obtained exact solutions in $O(n^6 \log^2 n)$ time. However they are limited to the case of planar sets.

Exact algorithms cannot be used in most practical applications where measurement errors and noise are present; furthermore, the high computational complexity of the exact algorithms make them impractical for use in real problems. For these reasons, approximate solutions for the case of point sets, both in 2-dimensional and in 3-dimensional space, have been considered [15].

In the field of computer vision, an efficient multi-resolution technique for comparing images using the Hausdorff distance has been presented in [29] where the space of possible transformations is limited to translations and scaling; in [48] the above technique is extended to affine transformations. Affine transformations are used in [24] for matching point sets. Other approaches to matching sets of segments in 3D space based on various techniques and metrics are given in [9], [24], [31]

The computation of the Hausdorff distance does not necessarily produce a one-to-one correspondence between the elements of the two sets; it may happen, in fact, that multiple elements in one set are associated with a single element of the other set. This is unlike most existing object recognition methods that give an explicit pairing.

Approaches have been proposed for the computation of the Hausdorff distance between sets of segments associated to secondary structures [10]. The standard Hausdorff distance provides a good metric over point sets but does not preserve the notion of relevant subsets like the segments. To keep information relative to the line segments in the definition of the distance function an alternative definition of the Hausdorff metric between sets of segments has been introduced in [23], together with efficient approximate algorithms for its computation Given two sets $A = \{a_1, a_2, \cdots, a_m\}$ and $B = \{b_1, b_2, \cdots, b_n\}$ of line segments a_i and b_j, the Segment Hausdorff distance $H_S(A, B)$ between A and B is:

$$H_S(A, B) = \max(h_S(A, B), h_S(B, A))),$$

where $h_S(A, B)$ is the one-way Segment Hausdorff distance given by:

$$h_S(A, B) = \max_{a_i \in A} \left(\min_{b_j \in B} H(\{a_i\}, \{b_j\}) \right)$$

The matching strategy is essentially an *alignment* that selects a few "representative" segments of the set A and computes a rigid transformation based on an hypothesized correspondence between the representative segments of A and a group of segments of B. It then verifies the hypothesis by computing the distance measure for such a transformation. The above steps are repeated for all possible groups of segments of A. More precisely, the algorithm looks for a rigid body transformation g (translation plus rotation) that minimizes the distance between two sets of segments, A and B and it consists of the following three main steps:

Step 1 determine a translation P;
Step 2 determine a rotation R;
STep 3 evaluate the distance between $g(A)$ and B, where g is the combined transformation.

The rigid body transformation is obtained by selecting three representatives for each of the two sets A and B that are affine independent elements.

First, a representative segment a for A is randomly chosen. This representative is paired with each element b of B. For each such pair (a, b), the translation P is defined by taking the mid-point a_m of a into the mid-point b_m of b. This choice of the translation minimizes the distance between the transformed segment $P(a)$ and b.

To define the rotation, two additional independent elements of A are needed. The second representative a' is chosen as the segment containing the point a'_f farthest from a_m. The third representative segment a'' is chosen so that it contains the point a''_d at maximum distance from the line $\overline{a_m a'_f}$. It is easy to see that the points a'_f and a''_d must each be an endpoint of some segment. The condition that is enforced is the affine independence of the three points a_m, a'_f and a''_d. These choices bind the error due to the approximation. The next step of the algorithm is to choose the segments b' and b'' of B in all the m^2 possible ways. For each b' and each endpoint of b', consider the rotation that has origin in a_m and that makes a'_f and a_m to become collinear with the endpoint of b'. Define R' as the one of the above rotations that minimizes the distance between a'_f and the endpoints of b'. Then define R'' to be the rotation about the axis $a_m a'_f$ that brings a''_d closest to an endpoint of b''. Apply the transformations $R'' R' P(A)$. Finally choose over all the triplets b, b' and b'' the transformation g that resulted in the smallest distance.

The time complexity of the overall algorithm is $O(mn^3 \log n)$. In fact, using the Hausdorff metric, the nearest neighbor query in a set of segments (to identify the segment of B "closest" to a segment of A), reduces to a nearest neighbor query among points in $\mathcal{R}^6$ that can be performed in optimal $O(\log n)$ time within a known error bound. It is shown that the error introduced with the approximation is within a bounded factor from optimal. This bound is the same as the bound obtained in [15] for the simpler case of point sets.

Experiments have been conducted on several proteins and the results were consistent with previous studies. As an example presented in [22], figure 6 shows the superposition of two sets of segments associated to proteins 1rpa and 1rpt, with very similar structures.

Fig. 6. The alignment of proteins 1rpa (red segments) and 1rpt (green segments)

7 Indexing Techniques

Indexing techniques, initially proposed in the field of computer vision by Wolfson et al., have found interesting applications in the area of bioinformatics. Indexing or geometric hashing provides a way to efficiently search a large database of proteins by storing redundant transformation invariant information about the proteins in a hash table, from which this information can be easily retrieved. The construction of the hash table, that constitutes the most complex part of the entire process, is done off-line at a preprocessing stage.

Indexing techniques have been applied to compare proteins at different levels of representations [2], [8], [13], [14], [32], [36], [55] (see also the chapter by H. Wolfson of this volume).

One major distinction of the comparison approaches is whether they are *order dependent* or *order independent*, in other words whether the use the order of the elements along the protein chain as a constraint in the correspondence process. Indexing techniques do not take into consideration the order of elements (either points or secondary structures) along the chain and therefore fall into the category of order-idenpendent methods.

For matching 3D point sets, quadruples of points are used to define reference frames or bases in which the coordinates of all other points remain invariant. Models are stored into the table by considering all possible combinations of quadruples of points as bases and using the invariant coordinates of the remaining points to index the table. At recognition time, if the correct quadruple of points is chosen from the image points, the candidate matches are efficiently retrieved from the corresponding entries of the hash table.

Here we concentrate on the application of hashing techniques at the secondary structure level involving transformation invariant properties of vectors associated to the secondary structures. The programs 3dSEARCH [51] and 3d-Lookup [26], based on hashing, compute geometric properties of pairs of secondary structures. They both construct the hash table by a procedure that consists of the following steps for the insertion of a protein in the database:

Step 1. For each pair of vectors of the protein, compute a reference frame or coordinate system identified by the two endpoints of one vector and by the orientation of the other vector.

Step 2. For each remaining vector in the protein, compute its coordinates in the reference frame defined in the previous step 1.

Step 3. The coordinates are quantized into fixed size interval and used to access the entry of the table corresponding to those coordinates where the following pair of information is stored: 1) name of the protein that hashed into it; 2) identifiers of the two vectors used as reference frame.

Once the hash table is built, each secondary structure vector from the given query structure is simultaneously compared to the entire library of target structures by simply indexing into this table. Thus, to compare a query protein to all target proteins the above step 1 and 2 are repeated for the query protein. Step 3 is replaced by the following:

Step 3' The coordinates are quantized into fixed size interval and used to access the entry of a 3d table corresponding to those coordinates where a vote is cast to every pair (protein name, two vector identifiers) present at that entry.

At the end of the process the proteins in the table which obtained the most votes are the candidates for matching.

A recently proposed approach [21] considers triplets of secondary structures rather than pairs to build the hash table. The three dihedral angles associated to all triplets of secondary structures are used to index a hash table. Let (s_i, s_j, s_k) be a triplet of segments, where s corresponds either to an α-helix or a β-strand. Let α_{sr} be the dihedral angle formed by two segments s and r. The dihedral angle between two segments is the angle formed by the two planes perpendicular to the straight lines containing the segments themselves and therefore is defined in the range $[0, 180]$.

The triplet of segments (s_i, s_j, s_k) is then described by the three angles $(\alpha_{ij}, \alpha_{jk}, \alpha_{ki})$. The three angles, quantized into uniform intervals, provide three indeces for the table; a fourth index *triplet_type* is used to access the table; it depends on the types of the secondary structures in the triplets, whether all α helices, one α helix and two β strands and so forth. Thus a 4-dimensional table is constructed during the set up phase of the method. No explicit information is present in the tables about the order of the segments along the polypeptide chain.

Each entry of the table keeps a record of each triplet hashed into it. The record contains the following information: 1) the name of the protein, 2) the identifiers for the three vectors, 3) the pairwise distances between the three vectors. Such distances are used to filter incorrect hypotheses of associations in the matching process, because false results could be obtained based on angular information only. The distance is measured as the distance between the middle points of two segments.

The construction of the table is computation intensive; it requires $O(n^3)$ time, for n secondary structures. Once it is built it allows fast retrieval of candidate matches between the query protein and the proteins stored in the database. The space requirements for this approach may be high. However, the table is only partially occupied since the three angles are related by the triangular inequality.

The table can be queried to find similarities in the arrangement of the secondary structures of a query protein with the proteins stored in the database. A protein P is matched against the database of proteins by the following procedure:

Step 1 For each triplet (s_i, s_j, s_k) of secondary structures of P, compute the three angles (α, β, γ) and the three distances of the associated segments.

Step 2 Access the cell of the hash table indexed by $(\alpha, \beta, \gamma, triplet\_type)$ and tally a vote for each entry in the cell with similar distance values.

Step 3 Formulate and rank hypotheses of matching by determining the proteins with the highest number of votes.

The verification of the hypothesized matches may be performed by a pairwise comparison between the proteins, either at the level of secondary structures [23] or by extending the matching to residue level.

Once compiled, the table can be used for different types of comparisons, for instance for all-to-all structure comparison.

Experiments have been conducted that consisted in building the hash table for all proteins in the PDB (approx. 14.000) have shown that the approach is both robust and efficient. The construction of such a table for approximately 350 representative proteins from the PDB has led to interesting observations about the distribution of the angles of the secondary structures which deviate from the distribution of randomly chosen vectors more significantly than one would have expected [47].

8 Graph-Theoretic Approaches

Here we give a brief description of the use of graphs to represent protein structures and of the techniques used to identify common substructures within proteins or to search for specific patterns in databases of molecules. Early work on protein matching based on graph theory was done by Brint and Willett [7]. To compare two or more structures, the method generates a graph of correspondences where each node represents a pair of atoms, one for each protein, and an edge connects two nodes (a, b) and (a', b') if the difference of the distances $d(a, a')$ and $d(b, b')$ is below a certain tolerance. Problem 3 defined in section 3, that is finding a maximum-cardinality set of element pairs such that the distance between each pair is at most a given value δ, can then be restated as one of determining the maximal clique of a graph. Finding a clique in a graph is a well known NP-complete problem.

In [57] graph isomorphism is used to search for pharmacophoric 3D patterns in a database. The method of representation of a 3D structure is the connection table, which contains the list of all atoms of the structure together with the bond information that describes the way in which the atoms are linked together. The connection table is basically a graph where the nodes represent the individual atoms and the edges represent interatomic distances. The graph is complete in the sense that there is an edge connecting every pair of nodes. A variant of this graph is obtained by using angular information to label the edges. The angle can be either the torsion angle or the valence angle defined as the angle between three bonded atoms. The presence of a query substructure in a database can be tested by means of a subgraph isomorphism. Subgraph isormorphism is computationally expensive and therefore cannot be used to test all entries in the database. Thus a screening strategy is suggested to reduce the overall execution time by eliminating most of the entries of the database from consideration by subgraph isomorphism. This screening is analogous of an index technique in that it provides access to a small fraction of the database by a preprocessing operation that groups all elements with similar characteristics. An extension to flexible matching is provided as well.

The approach by Escalier at al. [12] finds the largest similar subsets of atoms by recursively building subsets of increasing sizes, combining two subsets of size k to build a subset of size $k + 1$. Two subsets can be combined to form a larger one if they differ in one element only and their inter-atomic distances are all below a given threshold. Thus, this problem is equivalent to a clique finding problem. A suitable tree data structure allows an efficient implementation of the merge operation. As stated by the authors, the approach is suitable for small (less than 30 atoms) molecules. For larger structures such as proteins, a brute force application of the algorithm may lead to unreasonable execution times. Heuristics have to be introduced to reduce the amount of computation. One such heuristic is to split the problem in two parts: the first is to identify local fragments and the second is to assemble them together.

A graph-based approach was used in [20], [41] to compare secondary structure motifs in proteins. Proteins and motifs are represented as labelled graphs with the nodes corresponding to the segments associated s to secondary structures and the arcs to the angular and spatial relations between segments. Subgraph isomorphism is used to identify common structural patterns in pairs of proteins or to search for motifs in the PDB. The Ullman's algorithm for subgraph isosmorphism [54] was found to be sufficiently fast for the search of small motifs from he PDB. Chains of lines segments, corresponding to either secondary structures or to linear representation of other fragments of the backbone have also been considered [1].

9 Integration of Methods for Protein Comparison Using Different Representations

Different representations offer richer source of information that can be used in the comparison. This approach has been already investigated in the literature resulting in effective tools. Alignment of superfamily members has been obtained through conservation of structural features such as solvent accessibility, hydrogen bonding and the presence of secondary structures [42], [45], [50], [53].

An hierarchical protein structure superposition using both secondary structures and backbone atoms was recently proposed by Singh and Brutlag [51]. The local alignment of secondary structures is obtained by a variation of the Smith-Waterman dynamic programming algorithm [52]. A score function is used in the dynamic programming to measure the degree of similarity between pairs of vectors (linear segments) and is an attribute that may be either orientation independent (like the angle between two vectors within the same protein) or orientation dependent (like the angle between two vectors corresponding to two structures each belonging to one protein of the pair). Other attributes may relate distances between segments within the same or different proteins. The expression for the score function S used in this approach, similar to that used by Gerstein and Levitt [18], is given by:

$$S = \frac{2M}{1+[d/d_0]^2} - M$$

where M is a weighting factor for the attribute being measured, d the attribute value and d_0 is the value at which the score should be 0. An important choice of a dynamic programming approach is how to assign gap penalties. For secondary structure alignment it may be appropriate to decide to introduce no penalty because often the deletion of a secondary structure is due to an incorrect assignment in the PDB or to a mutation that changed a single secondary structure element, say a strand, into two structures or converted a strand into a turn.

Once an initial superposition of the secondary structures has been obtained by the dynamic programming algorithm, it is refined by iteratively minimizing the RMSD between pairs of nearest atoms from the two proteins. An interesting feature of the approach is that it does not simply rely on RMSD for judging the quality of the alignment but it takes into consideration also the number of "well" aligned atoms. Well aligned atoms define the "core" of the proteins and are selected as follows. Pairs of atoms, one atom from each protein, are selected so that each atom of the pair is the nearest atom of the other atom of the pair in the other protein. Furthermore, to be included in the core, such pairs of atoms have to satisfy the co-linearity property i.e. if (i, j) and (h, k) are two such pairs and $i < h$ then it must be $j < k$. Thus this method is order-dependent, according to one major classification of protein comparison approaches. The last step of the algorithm is to try to improve on the superposition of the core atoms even at the cost of degrading the alignment of the rest of the atoms.

The algorithm is efficient in terms of computational complexity and spends most of the execution time on the secondary and atomic alignment and a small fraction on the alignment of the core structures.

10 Conclusions

The problem of protein comparison can be successfully approached by first considering the related geometric issues. In the paper, the power and limitations of the different algorithms for protein structure comparison have been reviewed and discussed. Most of them have already proved their utility in computer vision and image processing, as well as in robotics, astronomy and physics. Their use in molecular biology and bioinformatics opens new perspectives for developing integrated methods for protein comparison, classification and engineering. Even if the different methods have been introduced to be used within different applications (characterized by different requirements), they solve particular instances of a more general matching problem, as deeply investigated in the area of computational geometry. The variety of protein representations supports the reasoning both at the level of points (i.e. atomic level) and at level of segments (or secondary structures). Estimation of rigid transformations with different metrics is an important technique within the protein structure comparison algorithms. Moreover geometric indexing techniques prove their effectiveness in searching large protein databases. Finally, graph-theoretic protein modeling helps in designing algorithms for substructure identification and comparison.

From the current research it has been recognized that the combination of different methods and different protein representations may result in new and effective algorithms with decreased computational complexity and better speed. Solvent accessibility, hydrogen bonding and the presence of secondary structures can be considered together in the alignment of superfamilies, while a hierarchical protein structure superposition can be obtained using both secondary structures and backbones atoms. For a better characterization of the proteins functions and their evolutionary information, geometric reasoning should be coupled with some proper chemical consideration involving hydrophobicity, charge, etc. The goal is to use domain-specific information for allowing a better pruning of the possible association choices at a very early stage of the matching process.

11 Acknowledgements

Support for Guerra was provided in part by the Italian Ministry of University and Research under the National Project "Bioinformatics and Genomics Research", and by the Research Program of the University of Padova.

Support for Ferrari was provided in part by the Italian Ministry of University and Research and by the Research Program of the University of Padova.

References

1. Abagyan, R.A. , Maiorov, V.N. (1989). An Automatic Search for Similar Spatial Arrangements of α helices and β-strands in globular proteins. *Journal of Molecular Structural Dynamics*, **6**, 5, 1045-1060.
2. Alesker, A., Nussinov, R., Wolfson H.J. (1996). Detection of non-topological motifs in protein structures. *Protein Engineering*, **9**, 5, 1103-1119.
3. Alter, T. D. (1992). Robust and efficient 3D recognition by alignment. Technical Report AITR-1410, Massachusetts Institute of Technology, Artificial Intelligence Laboratory.
4. Arun, K.S., Huang,T.S., Blostein, S.D. (1987). Least-square fitting of two 3-D point sets. *IEEE Trans. on Pattern Analysis and Machine Intelligence*, **9**, 5, 698-700.
5. Branden, C., Tooze, J. (1999) *Introduction to Protein Structure*, (second edition), Garland.
6. Brown, N.P., Orengo C.A., Taylor W.R. (1996). A protein structure comparison methodology. *Comp. Chem.* , **27**, 359-380.
7. Brint, A.T., Willett, P. (1987). Algorithms for the identification of three-dimensional maximal common substructures. *J. Chem. Inform. Comput. Sci.*, **27**, 152-156.
8. Califano, A., Mohan, R. (1992). Multidimensional indexing for recognizing visual shapes. *IEEE Trans. on Pattern Analysis and machine Intelligence*, **16**, 4, 373-392.
9. Chen, H. H., and Huang, T. S. (1990) Matching 3d line segments with application to multiple-objects motion estimation. *IEEE Transactions on Pattern Analysis and Machine Intelligence*, **12**, 1002-1008.

10. Chew, P.L., Dor D.,pEfrat A., Kedem, K. (1995). Geometric pattern matching in *d*-dimensional space, *Algorithms-ESA '95*, 264-279.
11. Chew, L.P., Goodrich M.T., Huttenlocher,D.P., Kedem,K., Kleinberg, J.M., Kravets, D. (1997). Geometric pattern matching under Euclidean motion, *Computational Geometry. Theory and Applications*, **7**, 113-124.
12. Escalier V., Pothier, J., Soldano, H., Viari, A. (1998). Pairwise and Multiple Identification of three-dimensional common substructures in proteins. *J. of Computational Biology*, **5**, 41-56,
13. Fischer, D., Bachar, O., Nussinov, R., and Wolfson, H. (1992). An efficient automated computer vision based technique for detection of three dimensional structural motifs in proteins. *J. Biomol. Struct. Dyn.*, **9**, 769-789.
14. Fischer, D., Tsai, C.J., Nussinov, R., and Wolfson, H. (1995). A 3D sequence-independent representation of the protein data bank. *Protein Engineering*, **8**, 981-997.
15. Goodrich, M.T., Mitchell, J.S.B., Orletsky, M.W. (1999) Practical Methods for Approximate Geometric Pattern Matching Under Rigid Motion. *IEEE Trans. Pattern Analysis and Machine Intelligence*, **21**, 4, 371-379.
16. A. Glassner, (editor), *Graphics Gems*, Academic Press, 1990.
17. Gerstein, M. (1992). A Resolution-Sensitive Procedure for Comparing Protein Surfaces and its Application to the Comparison of Antigen-Combining Sites. *Acta Cryst.*, **A8**, 271-276.
18. Gerstein, M., Levitt, M. (1998). Comprehensive Assessment of automatic structural alignment against a manual standard, the scop classification of proteins. *Protein Science*, **7**, 445-456.
19. Golub, G.H., Van Loan C.F. , (1996) *Matrix Computation*, Johns Hopkins University Press.
20. Grindley, H.M., Artymiuk, P.J., Rice, D.W., and Willett, P. (1993). Identification of tertiary structure resemblance in proteins using a maximal common subgraph isomorphism algorithm. *J. Mol. Biol.*, **229**, 707-721.
21. Guerra, C., Lonardi, S., Zanotti, G., (2002). Analysis of proteins secondary structures using indexing techniques. *IEEE Proc First Int. Symposium on 3D Data Processing Visualization and Transmission*, 812-821.
22. C. Guerra, V. Pascucci. (1999) On matching sets of 3D segments", *Proceedings of SPIE Vision Geometry VIII*, 157-167.
23. Guerra, C., Pascucci, V. (1999) 3D segment matching using the Hausdorff distance. *Proceedings of the IEEE Conference on Image Processing and its Applications, IPA99*, 18-22.
24. Hagedoorn, M., Veltkamp, R. C. (1999). Reliable and efficient pattern matching using affine invariant metric. *Int. J. of Computer Vision*, **31**, (2/3), 203-225.
25. B. Horn, H. Hilden, S. Negahdaripour (1988). Closed-form solution of absolute orientation using orthonormal matrices. *J. Opt. Soc. Am.*, **5**, 1127-1135.
26. Holm, L., Sander, C. (1995). 3D-Lookup: Fast protein structure database searches at 90% reliability. *Proc. Third Int. Conf. on Intell. Sys. for Mol. Biol.*, Menlo Park, 179-187.
27. Holm, L., Sander, C. (1996). Mapping the protein universe. Science, *Science*, **273**, 595-602.
28. B. Horn, (1987). Closed-form solution of absolute orientation using unit quaternions. *J. Opt. Soc. Am.*, **4**, 629-642.

29. Huttenlocher, D. P., Klanderman, G. A., and Rucklidge, W. J. (1993) Comparing images using the Hausdorff distance. *IEEE Transactions on Pattern Analysis and Machine Intelligence*, **15**, 9, 850-863.

30. Kabsch, W., Sander, C., (1983). Dictionary of protein secondary structure: pattern recognition of hydrogen-bonded and geometrical features. *Biopolymers*, **22**, 2577-2637.

31. Kanmgar-Parsi, B., and Kamgamr-Parsi, B. (1997) Matching sets of 3d line segments with application to polygonal arc matching. *IEEE Transactions on Pattern Analysis and Machine Intelligence*, **19**, 10, 1090-1099.

32. Y. Lamdan, J. T. Schwartz, H. J. Wolfson. (1990). Affine invariant model-based object recognition, *IEEE Trans. on Robotics and Automation*, 578-589.

33. Lancia, G., Carr, R., Walenz, B., Istrail, S. (2001). Optimal PDB structure alignments: A Branch-and-Cut algorithm for the maximum contact map overlap problem. *Proc. 5th ACM REsearch in COMputational Biology*, 193-202.

34. Laskowski R.A., MacArthur M.W., Moss D.S., Thornton J.M. (1992). Stereochemical duality of protein structure coordinates. *Proteins*, **12**, 345-364.

35. Laskowski R.A., MacArthur M.W., Moss D.S., Thornton J.M. (1993). PROCHECK: a program to check the stereochemical quality of protein structures. *J. Appl. Cryst.* , **26**, 283-291.

36. Leibowitz, N., Fligelman, Z.Y., Nussinov, R., Wolfson, H.J. (1999). Multiple structural alignment and core detection for geometric hashing. *Proc. ISMB99*, Heidelberg, Germany, 169-177.

37. Lemmen, C., Lengauer, T. (2000). Computational methods for the structural alignment of molecules. *J. of Computer-Aided Molecular Design*, **14**, 215-232.

38. Lesk, A. M. (1991).*Protein architecture: a practical approach*. Oxford Univ. Press, Oxford.

39. Lesk, A. (1994). Computational Molecular Biology. *Encyclopedia of Computer Science and Technology*, **31**, Marcel Dekker, Inc..

40. Eggert,D.W., Lorusso, A., Fisher, R.B. (1997). Estimating 3-D rigid body transformations: a comparison of four major algorithms, *Mach. Vis. and Applic.*, **9**, 272-290.

41. Mitchell, E.M., Artymiuk, P.J., Rice, D.W. Willett, P. (1989) Use of Techniques derived from graph theory to compare secondary structures motifs in proteins. *J. Molecular Biology*, **212**, 151-166.

42. Mizuguchi, K., Deane, C.M., Blundell, T.L., Johnson, M.S. and Overington,J.P. (1998). JOY: protein sequence-structure representation and analysis. *Bioinformatics*, **14**, 617-623.

43. Murray, R.M, Li, X., Sastry, S.S., (1994). *A Mathematical Introduction to Robotic Manipulation*, CRC Press.

44. Murzin, A.G., Brenner, S.E., Hubbard, T., Chothia, C. (1995) SCOP: a structural classification of proteins database for the investigation of sequences and structures. *J. Molecular Biology*, **247**, 536-540.

45. J. P. Overington, Z. Y. Zhu, A. Sali, M. S. Johnson, R. Sowdhamini, G. V. Louie, and T. L. Blundell. (1993). Molecular recognition in protein families: a database of aligned three-dimensional structures of related proteins. *Biochem. Soc. Trans.*, **21** , 3, 597-604.

46. Pennec, X., Ayache, N. (1998) A geometric algorithm to find small but highly similar 3D substructures in proteins. *Bioinformatics*, **14**, 6, 516-522.

47. Platt, D.E., Guerra, C., Rigoutos, I., Zanotti, G. (2002). Global secondary structure packing angle bias in proteins. Manuscript.

48. Rucklidge, W. J. (1997) Efficiently locating objects using the Hausdorff distance. *International Journal of Computer Vision*, **24**, 3, 251-270.
49. Sabata, B., Aggarwal, J.K. (1991). Estimatiom of motion from a pair of range images: a review. *Computer Vision, Graphics and Image Processing: Image Understanding*, **54**, 3, 309-324.
50. Sali A., Blundell, T.L. (1990). Definition of a General Topological Equivalence in protein structures: a procedure involving comparisons of properties and relationships through simulated annealing and dynamic programming, *Journal of Molecular Biology*, **212**, 403-428.
51. Singh,A.P., Brutlag, D.L., (1997). Hierarchical protein structure superposition using both secondary structures and atomic representations. *Proc. Fifth Int. Conf. on Intell. Sys. for Mol. Biol.*, Menlo Park, 284-293.
52. Smith, T.F., Waterman, M.S., (1981). Identification of common molecular subsequences. *J. Mol. Biol.*, **147**, 195-197.
53. Sowdhamini R., Burke D.F, Huang J-F, Mizuguchi, K. Nagarajaram H.A., Srinivasan N., Steward RE. and Blundell T.L. (1998). CAMPASS: A database of structurally aligned protein superfamilies. *Structure*, **6**, 9, 1087-1094.
54. Ullman, J.R. (1995) *J. Assoc. Comp.*, **23**, 31-42.
55. Verbitsky, G. Nussinov, R., Wolfson H.J. (1998). Structural comparisons allowing hinge bendings, swiveling motions. *Proteins*, **34**, 232-254.
56. Walker M.W, Shao L., Voltz R.A., Estimating 3-D Location Parameters Using Dual Number Quaternions, *CVGIP:Image Understanding*, **54**, 3, 1991, 358-367
57. Willett, P. (1995) Searching fore pharmacophoric patterns in databases of three-dimensional chemical structures. *J. of Molecular Recognition*, **8**, 290-303.

Identifying Flat Regions and Slabs in Protein Structures

Mary Ellen Bock[1] and Concettina Guerra[2]

[1] Department of Statistics, Purdue University, West Lafayette, 47907 IN, USA
mbock@stat.purdue.edu
[2] Dipartimento Elettronica e Informatica, Via Gradenigo 6/A, Padova, Italy or
Department of Computer Science, Purdue University,
West Lafayette, 47907 IN, USA
guerra@dei.unipd.it

Summary. Motivated by the problem of identifying flat regions in three dimensional protein structures, we provide a new geometric approach for the extraction of planar surfaces that compares favorably with existing approaches developed for computer vision. Preliminary results obtained on proteins from different structural classes are given.

Index Terms - Plane detection - Hough transform - Width - Protein Structure analysis

1 Introduction

When we examine the 3D tangle of a protein string we sometimes observe flat or planar regions. Certain large segments or groups of amino acids lie roughly in a plane or very narrow slab. The presence of these segments or groups of amminoacids in a very narrow slab can be very useful for confirming or suggesting potential secondary structures such as a β-strand. The episodic occurence of this phenomenon motivated an interest in general algorithms that quickly find flat regions in 3D point sets.

We formulate the problem of identifying flat regions as follows. Given a set Γ of n points in 3D space and a non-negative constant ϵ, determine the plane that is at a distance at most ϵ from the maximal number of points of Γ.

In other words, we want to determine a pair of parallel planes (slab) a distance of 2ϵ apart enclosing a maximal number of points of Γ. The pair of planes is

C. Guerra, S. Istrail (Eds.): Protein Structure Analysis and Design, LNBI 2666, pp. 83-97, 2003.
© Springer-Verlag Berlin Heidelberg 2003

not necessarily unique. For the extraction of multiple planar regions from the input data, the same problem is solved repeatedly after the removal of the points that are found to belong to the best slab.

We present a new method for the detection of planar regions in point sets; we also give some heuristics that can help speed-up the process while maintaining a good quality for the solution. We show that our geometric method has many advantages over the Hough transform, a popular technique to extract parametric shapes, such as lines, planes, and circles from images.

Our detection algorithm solves the problem as a maximum coverage problem in the parameter space. A subset of three points of Γ determines a region in the parameter space containing the parameters of all planes within ϵ distance from the three points. Consider the arrangement in the parameter space formed by the regions determined by all subsets of three points of Γ. Define the *depth* of a point in the parameter space as the number of regions that contain the point. This is in fact the number of points of Γ within ϵ distance from the corresponding plane. The point of maximum depth is sought among the vertices of the arrangement. Each such vertex gives the parameters of a plane at distance either $+\epsilon$ or $-\epsilon$ from each of the three points.

The search for a vertex of maximum depth can be done by simply counting for a plane corresponding to a vertex of the arrangement the number of points of Γ within ϵ distance from it and then choosing the plane that has the largest count. This computation requires $O(n)$ time for each plane. A more efficient solution to this problem is also presented that avoids the above computation for all planes, by exploiting the structure of the arrangement of the regions in the parameter space. The algorithm has a $O(n^3 \log n)$ time complexity.

The problem of detecting geometric primitives is a fundamental one in computer vision; there it is often solved by the Hough transform, a method that transforms the extraction problem into an intersection problem in the parameter space, that is is quantized in an accumulator array.

The most common application of the Hough transform in image processing is for line detection. The method has been also applied to circle and ellipse detection, and to the extraction of planes from range images. If storage space is a concern this approach can be used only for parametric shapes with few parameters, since the accumulator array dimensionality grows with the number of the parameters.

The plane extraction problem is related to a well-known problem in computational geometry, the *width* problem, that is the determination of a pair of parallel planes of minimum distance enclosing all points of a given set. Houle and Toussain [9] give algorithms that produce the exact width of a point set and run in $O(n \log n)$ time in $\Re^2$ and in $O(n^2)$ in $\Re^3$. An efficient approximation to the width determination is given in [4], that can be used to perform the metrology primitive of "flatness". The requirement that all points are within a single slab formed by the two parallel planes makes these approaches not suitable for applications that require the extraction of multiple slabs with a predefined width.

2 A Geometric Algorithm

Given a set Γ of n points in 3D-space and a non-negative real value ϵ, we want to determine a plane P in $\Re^3$ that is at a distance at most ϵ from the largest number of points of Γ.

Let the equation of plane P in $\Re^3$ be

$$ax + by + cz + d = 0$$

where (a,b,c,d) is the 4-dimensional vector of parameters of P satisfying $a^2 + b^2 + c^2 = 1$.

We solve the 3D plane detection problem by solving an intersection problem in the $\Re^4$ parameter space. To each three-dimensional point $p_i \in \Gamma$ we associate a region $R_i \in \Re^4$ (as shown below). Each of the four-dimensional points in R_i corresponds to the parameters of a plane in $\Re^3$ at a distance at most ϵ from p_i. If (a, b, c, d) is a four-dimensional point in the intersection of the two regions R_i, R_j associated to the three-dimensional points p_i, p_j, then the corresponding plane in $\Re^3$ given by the equation $ax + by + cz + d = 0$ is at a distance at most ϵ from both points p_i and p_j in $\Re^3$. We define the *depth* of a point in $\Re^4$ to be the number of regions R_i that contain it and look for a point in $\Re^4$ with maximal depth. (The point is not unique.) Another way of looking at this problem is to view it as a coloring problem. Each region is colored by a different color; we say that a point is colored by a given color if it belongs to a region with that color. The objective is to find a point that is covered by the maximum number of different colors.

The region R_i in $\Re^4$ associated to point $p_i = (x_i, y_i, z_i)$ in $\Re^3$ can be obtained as follows. First if a plane P' in $\Re^3$ has parameters (a', b', c', d') and if the plane contains p_i then it satisfies:

$$a'x_i + b'y_i + c'z_i + d' = 0.$$

The above equation also describes in $\Re^4$ a hyperplane $Q_i{}^0$ through the origin with coefficient $(x_i, y_i, z_i, 1, 0)$. Let $d_{p_i,P}$ be the distance in $\Re^3$ of point p_i from the plane P with parameters (a,b,c,d). That is,

$$d_{p_i,P} = |ax_i + by_i + cz_i + d| \tag{1}$$

since $a^2 + b^2 + c^2 = 1$.

The region R_i in $\Re^4$ consists of points (a, b, c, d) satisfying:

$$\begin{cases} ax_i + by_i + cz_i + d + \epsilon \geq 0 \\ ax_i + by_i + cz_i + d - \epsilon \leq 0 \\ a^2 + b^2 + c^2 = 1 \end{cases} \tag{2}$$

R_i has boundaries given by the two hyperplanes $Q_i^{+\epsilon}$ and $Q_i^{-\epsilon}$ with respective coefficients $(x_i, y_i, z_i, 1, +\epsilon)$ and $(x_i, y_i, z_i, 1, -\epsilon)$ and the hypercylinder

$$a^2 + b^2 + c^2 = 1 \tag{3}$$

The points (a, b, c, d) in $Q_i^{+\epsilon}$ satisfy:

$$ax_i + by_i + cz_i + d + \epsilon = 0$$

The points (a, b, c, d) in $Q_i^{-\epsilon}$ satisfy:

$$ax_i + by_i + cz_i + d - \epsilon = 0$$

Note that the two above hyperplanes $Q_i^{+\epsilon}$ and $Q_i^{-\epsilon}$ are parallel to the hyperplane Q_i^0 and enclose it. Furthermore, both $Q_i^{+\epsilon}$ and $Q_i^{-\epsilon}$ have distance from Q_i^0 given by: $\frac{\epsilon}{\sqrt{x_i^1 + y_i^2 + z_i^2 + 1}}$.

We consider the arrangement of all regions R_i in $\Re^4$. It is easy to see that region R_i cannot be contained in R_j, $i \neq j$. This is because there always exists a plane within ϵ distance of p_i that is farther than ϵ from p_j. Furthermore, the intersection of any three regions $R_i \cap R_j \cap R_k$, $i, j, k = 1, ..n$, is non empty. In fact the intersection contains at least the point corresponding to the parameters of the plane through the three points. The boundaries of three regions intersect in at most 8 *vertices*, each obtained by intersecting three hyperplanes, one for each region, and the hypercylinder. As far as the depth information is concerned only the boundaries of the regions are relevant. Thus we look for a point with maximum depth among the vertices of the arrangement.

Consider the 8 vertices in $\Re^4$ determined by R_i, R_j, and R_k. They correspond to 8 candidate solution planes in $\Re^3$. However the triple of points p_i, p_j, and p_k in $\Re^3$ corresponding to the three regions does not lie in any of the candidates planes. Each candidate plane can be described as the middle plane of a pair of parallel planes a distance of 2ϵ apart that encloses the maximum number of points of Γ. This middle plane is parallel to the two members of the pair and halfway between them. In the first two of the 8 cases, the triple of points lies on one of the members of the pair, i.e. the first candidate optimal plane is at distance ϵ and the second at distance $-\epsilon$ from each of the three points p_i, p_j, and p_k. In the last 6 cases, two points of the original triple lie on one member of the pair of enclosing planes while the third point lies on the other member of the pair. In other words, by choosing an unordered pair of points from the triple (and there are three possible choices), two planes are determined that are between the pair of points and the remaining member of the triple and that are at distance ϵ (in absolute value) from each of the three points.

The problem of finding a vertex of maximum depth has a simple solution in the original space. For each vertex count the number of points in $\Re^3$ within ϵ distance from the corresponding plane. Then select the vertex with maximum count. Since there are $O(n^3)$ vertices in the arrangement this procedure takes $O(n^4)$ time. In the next section we present a more efficient algorithm that exploits the structure of the arrangement of the regions and achieves $O(n^3 \log n)$ time complexity. We have implemented the algorithm also using random sampling, that is the selection of a small number of triplets of points from the input data set. The results are satisfactory in most cases; we do not however discuss them here for lack of space.

For the template matching cost function, methods which generate only the first two cases above for the candidate planes, i.e. the triple of points always lies in one of the two planes which form the enclosing pair, can miss 25% of the points as seen in example 2.1. Again for the template matching cost function, methods which always place each triplet of points in the middle plane (and form the potentially optimal pair of planes by creating parallel planes on either side of the middle plane a distance of ϵ away) can miss (25%) of the points as seen in example 2.2.

Example 2.1. Consider the following four points: $P_1(0,.75,0)$, $P_2(0, -.75,0)$, $P_3(1,0, .75)$, and $P_4(1, 0, -.75)$. Let $\epsilon = 0.5$. The pair of parallel planes, one with coefficients $(1,0,0,-1)$ containing P_3 and P_4 and the second parallel to it and containing P_1, encloses all four points (in fact P_2 belongs to the second plane). The solution generated by our method is the middle plane of this pair with coefficient $(1,0,0,-.5)$ and it is within distance 0.5 from all four points.

On the other hand, restricting one of the two planes which form the pair to always contain three of the four points misses one point. The middle plane of such a pair has always distance greater or equal to 0.7 from the remaining point. See for example, the plane $(0.6,0,-0.8,-0.5)$ which is middle plane of a pair with one member containing P_1, P_2 and P_3. Thus the maximal number of points within distance 0.5 from such planes is 3.

Example 2.2. Consider the same four points as in the previous example: $P_1(0,.75,0)$, $P_2(0, -.75,0)$, $P_3(1,0, .75)$, and $P_4(1, 0, -.75)$. Let $\epsilon = 0.5$. The planes containing three of the four points always have the distance 1.2 from the remaining point. Again the maximal number of points within distance 0.5 from such planes is 3.

An interesting feature of the algorithm is that its time complexity is independent of ϵ. Therefore it is the same even when high precision is required (very small values of ϵ). This is unlike the methods based on the Hough transform for which the complexity in time and memory increases with the degree of required precision.

3 An Improved Geometric Algorithm

In this section we present an efficient algorithm for finding a vertex of maximum depth in the arrangement of the R_i regions. We recall that a vertex is determined by the intersection of three hyperplanes at the boundary of three distinct regions R_i and the hypercylinder.

We first observe that the vertices of the arrangement lie on the $O(n^2)$ ellipses e_{ij} formed by the intersection of two hyperplanes $Q_i{}^{+\epsilon}$ and $Q_j{}^{+\epsilon}$, $i, j = 1, ..., n$, and the hypercylinder. More precisely, of the 8 vertices generated by R_i, R_j and R_k 4 belong to the ellipse e_{ij} and 4 to the ellipse e_{ik}. (or to the ellipse e_{jk}) On e_{ij} (e_{ik}) one pair of vertices is obtained by intersecting e_{ij} with $Q_k{}^{+\epsilon}$ and the second pair with $Q_k{}^{-\epsilon}$ ($Q_j{}^{+\epsilon}$ and $Q_j{}^{-\epsilon}$). Each pair of vertices on e_{ij} defines an arc consisting of points in $R_i \cap R_j \cap R_k$. Consider all arcs on e_{ij} obtained for all different values of k. The depth of a vertex is clearly given by 2 plus the number of arcs it belongs to.

Thus the problem of finding the vertex of maximum depth becomes that of computing the maximum overlap in a set of arcs. This problem can be solved using standard geometric techniques in $O(k \log k)$ time for k arcs [5]. Since this step is repeated for each ellipse corresponding to a pair of regions $R_i R_j$, the algorithm has an overall $O(n^3 \log n)$ time complexity.

The equation of the ellipse e_{ij} can be obtained as follows. Let

$$ax_i + by_i + cz_i + d + \epsilon = 0$$
$$ax_j + by_j + cz_j + d + \epsilon = 0$$
$$a^2 + b^2 + c^2 = 1$$

be the equations of the hyperplanes $Q_i{}^{+\epsilon}$, $Q_j{}^{+\epsilon}$ and the hypercylinder, respectively.

The intersection of the two hyperplanes in $\Re^4$ can be represented by:

$$a = su_a + tv_a$$
$$b = su_b + tv_b$$
$$c = su_a + tv_c$$
$$d = -\epsilon + su_d + tv_d$$

where s and t are parameters and $u = (u_a, u_b, u_c, u_d)$ and $v = (v_a, v_b, v_c, v_d)$ are orthogonal unit vectors. We can choose the vector v so that $v_d = 0$. The intersection of the above plane with the hypercylinder $a^2 + b^2 + c^2 = 1$ is given by:

$$(u_a^2 + u_b^2 + u_c^2)s^2 + (u_a v_a + u_b v_b + u_c v_c)s \cdot t + (v_a^2 + v_b^2 + v_c^2)t^2 = 1$$

Since $v_d = 0$ and $u \perp v$ the equation above is rewritten as:

$$\alpha s^2 + t^2 = 1$$

where $\alpha = 1 - u_d^2 \leq 1$. The above is the equation of the ellipse on the plane.

4 Hough Transform

The Hough transform is a powerful technique developed in the area of image processing for the detection of parametric curves in an image. It was originally proposed for line detection from gray-level images [8] and later extended to other parametric curves (circles, ellipses, etc.) [3] and to the recognition of arbitrary shapes. There exist several variants of the basic Hough transform; for recent surveys on the subject see [13], [16].

The standard formulation of the Hough transform for line detection is as follows. Suppose we are given a set of image points - for instance edge points or some other local feature points - and we want to determine subsets of them lying on straight lines. The Hough transform associates an image point with a line in the parameter space. The method is based on the property that collinear points are transformed into lines intersecting at the same point in the parameter space. The problem is then solved as an intersection problem in the parameter space, that is quantized in an array of cells, called the *accumulator array*. Specifically, the Hough transform consists of the following steps.

step 1. For each image point, determine the parameters of all lines through it and accumulate the corresponding cells in the parameter space.

steo 2. Find the maximum (or the local maxima) in the accumulator array. The corresponding parameters describe the detected line (lines).

An important property of the Hough transform is its insensitivity to noise and to missing parts of lines. The disadvantages are its high computational and memory requirements which increase with the resolution of the accumulator array. Furthermore, since the array dimensionality grows with the number of the curve parameters, this approach is applicable only to curves with few parameters. To overcome these limitations, many approaches have been proposed for image and vision vision applications, including probabilistic [11], randomized [12] and fast [15] Hough approaches.

Some of these approaches formulate the Hough transform as a many-to-one mapping from the original set of points into the parameter space, instead of a one-to-many mapping as in the standard formulation. For example, in the case of line detection that means determining for any two points of the set the line passing through them and incrementing the corresponding entry of the array in the parameter space. In general, for the detection of a curve represented by m parameter equations, the original approach maps one point into a $m-1$ hypersurface into the parameter space, while the second approach maps m points which define a curve into one point of the parameter space. The second approach is generally used in conjunction with *random sampling* [6], that is the random selection of few subsets of n points from the original input to reduce storage and computation time.

The Hough transform is easily extended to the detection of planes in a set Γ of points in 3D space. A plane is parametrized by the following expression, containing three parameters α, β and d:

$$\cos \alpha \cos \beta x + \cos \alpha \sin \beta y + \sin \alpha z + d = 0 \qquad (4)$$

Note that if we let

$$a = \cos \alpha \cos \beta \qquad (5)$$
$$b = \cos \alpha \sin \beta \qquad (6)$$
$$c = \sin \alpha$$

we have $a^2 + b^2 + c^2 = 1$, that is the same constraint used in the previous section.

The parameter space is quantized into $h \times k \times t$ cells that form the accumulator array A; each entry of the array corresponds to a triple (α, β, d). In other words the parameters α and β are discretized into h and k values, respectively, taken at regular intervals in the range $[0, \pi]$. The parameter d, which represents the distance of the plane from the origin, is discretized in t values; the range of t values is chosen according to the point set.

The plane detection algorithm proceeds in two steps. Step 1 constructs the accumulator array. It does so by computing for each point of the set Γ and for each pair (α, β) the d value from equation 4 and incrementing the corresponding entry in the accumulator array.

Step 2 of the algorithm determines the maximum in the accumulator array A, that corresponds to the detected plane.

For a set with n points, the Hough transform requires $O(n \times h \times k)$ time for step 1 and $O(h \times k \times t)$ for step 2.

Alternatively, the Hough transform for plane detection can be formulated as a many-to-one mapping from the set of points into the parameter space. For each triple of input points the plane through them is determined and only the corresponding entry in the accumulator array is incremented. For this alternative approach, the time for the construction of the accumulator array is $O(n^3)$, while the time for searching the array to determine its maximum value is still dependent on the quantization of the array, that is $O(h \times k \times t)$. As pointed out in the section 2, this method may fail to detect the best plane, defined as the plane within a given distance from the maximal number of points.

5 Performances of the Two Algorithms

The time and space complexity of the Hough transform depends on the number of points as well as on the quantization of the parameter space. For a coarse quantization of the accumulator array the algorithm is very efficient. There are however many applications requiring high resolution and therefore a fine discretization of the parameter space. The time complexity of the geometric algorithm is instead dependent only on the number n of input points and independent of ϵ.

We have compared the two approaches on several input sets of varying sizes and for different values of ϵ. To compare the two approaches in terms of time, we have first experimentally determined the quantization of the parameter space in the Hough approach that allows us to obtain results of the same quality as the geometric algorithm. The quality of the result is expressed by the number of points detected within ϵ distance from the best plane. In all experiments we observed that, with a proper choice of the quantization of the accumulator array, the standard Hough transform obtained the same maximum value as the geometric algorithm.

In Tables 1 and 2 the execution times of the two algorithms for small sets of points are reported. The sets of points consist of fragments of varying sizes of the backbone of the proteins 1KDU and 3PGM. For each of the two ϵ values, the required quantization step in the accumulator array for each of the three parameters α_{step}, β_{step} and d_{step} is shown along with the overall execution time T_{Hough}. The execution time T_{geom} of the geometric algorithm, which is the same for all ϵ values, is shown in column 2.

In all our tests the geometrical algorithm outperformed the Hough transform. The time for the Hough transform is dominated by the complexity of the second phase, i.e. finding the maximum in the accumulator array. As the size of the set increases, the memory requirements of the Hough transform increase significantly, and so does its time complexity. This is also because in our examples the space occupancy of input points (in terms of the diameter) increases with n requiring larger values for the parameter d.

There can be cases (depending on ϵ) where the Hough transform runs faster. However, given a set of points it seems that is always possible to find a value of ϵ for which the geometrical algorithm outperforms the Hough transform.

1KDU		$\epsilon = 1$				$\epsilon = 2$			
Num. of points	T_{geom}	T_{Hough}	α_{step}	β_{step}	d_{step}	T_{Hough}	α_{step}	β_{step}	d_{step}
20	0.1	17.7	2	2	0.2	3.9	3.5	3.5	0.7
25	0.2	17.7	2	2	0.2	3.5	3.5	3.5	0.7
30	0.5	17.7	2	2	0.2	5.7	3.5	3.5	0.7
35	0.9	21.0	2	2	0.2	4.3	3.5	3.5	0.7
40	1.5	36.0	1.8	1.8	0.2	10.3	3	3	0.7
45	3.1	>3000	1.2	1.2	0.2	21.5	2.5	2.5	0.7

Table 1. Execution times of the two algorithms for fragments of the protein 1KDU.

3PGM		$\epsilon = 1$				$\epsilon = 2$			
Num. of points	T_{geom}	T_{Hough}	α_{step}	β_{step}	d_{step}	T_{Hough}	α_{step}	β_{step}	d_{step}
20	0.1	0.3	9	9	0.5	0.4	9.9	9.9	0.6
30	0.6	1.4	7.2	7.2	0.3	0.5	9	9	0.5
40	1.8	10.6	3	3	0.2	2.8	5.6	5.6	0.3
50	4.4	36.2	1.8	1.8	0.2	27.9	2	2	0.5

Table 2. Execution times of the two algorithms for fragments of the protein 3PGM

6 Plane Detection in Proteins

One application of the above method is in improvement of protein secondary structures assignment from atomic coordinates. This is an important task because it is often a preliminary step in the analysis of the protein structures; for instance secondary structures can be used in protein structural comparison to identify largest common substructures or to search for structural motifs in the Protein data Bank (PDB) [1]. The task of assigning secondary structures may be based on different criteria which may lead to slightly different solutions; they include hydrogen bond pattern, backbone torsion angles, and inter-Ca atoms distances [7], [10], [14]. A recent comparison of three different procedures for the assignment of secondary structures shows that they agree in only about 60 % of the sequence sites in several proteins [2].

The algorithm we have presented may be used in identifying wrong assignments or, more importantly, in deriving properties that can be useful in secondary structure prediction.

We have determined the widths of the β-strands and α-helices of a small set of representative protein structures from the PDB database. The set includes the following protein structures:

1aei 1eur 1emd 1gga 1ovi 1pal 1tie 1acp 1azu 1bgt 1cbn 1cdg 1cll 1dat 1dfn 1fha 1gca 1gdl 1gpv 1hcv 1hpi 1ifb 1igf 1igm 1paz 1pml 1pyk 1rbp 1rpa 1rpr 1rpt 1sbp 1stp 1svc 1tim 1ttf 1utg 2mhr 2aza 2bop 2bpa 2cpl 2ctc 2cyh 2dri 2hla 2hnq 2liv 2paz 2pf2 2plh 2plt 2spt 2taa 3por 8paz

where the four-symbol strings represent the identifiers in the PDB.

Table 4 shows the different width values obtained for any given length of the β-strands. Only the Ca atoms are included in the analysis. The lengths of the β-strands considered range from 6 to 23 residues. Table 3 has the same meaning except that the α-helices are considered. The lengths of the helices range from 6 to 23 residues. Short helices and strands have not been considered in this analysis. Repeated secondary structures in different chains of a protein have been counted only once.

In table 3 there are only two values that are far from the average width value. One such value (6.9) corresponds to the α-helix in figure 4. The picture clearly indicates a bad secondary structure assignment. The other value corresponds to the α-helix in figure 3. On both α-helices the geometric algorithm that finds the maximum number of points within ϵ distance from a plane was applied. For the α-helix of figure 3 the algorithm with a value of ϵ equal to the average width of all helices correctly identified the last fragment of the secondary structure as being outside of the slab of average width. For the α-helix of figure 4 the algorithm found one "outlier", i.e. the residue with a darker color in the figure. Figure 5 shows a β-strand with a width value above average, and again the geometric algorithm found the extreme residues outside the best slab of average width.

We believe that this analysis can be of help in the study of the relations between residue-related parameters (such as the volume, etc.) and the 3D geometry of the secondary structures that can help predict the secondary structures from the sequence of amino-acids.

We have also applied the algorithm to the detection of slabs of given widths enclosing the maximum numbe of points to several proteins from different structural classes. In figures 6 and 7 we show the results for 6 proteins from the PDB; in all cases we have considered the Ca atoms only.

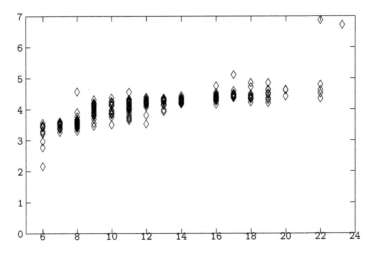

Fig. 1. Widths of helices. (helix length on horizontal axis)

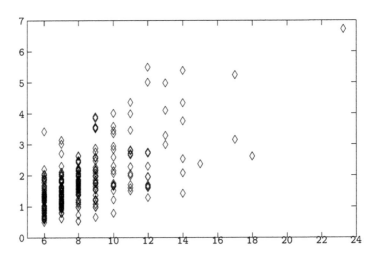

Fig. 2. Widths of β-strands. (strand length on horizontal axis)

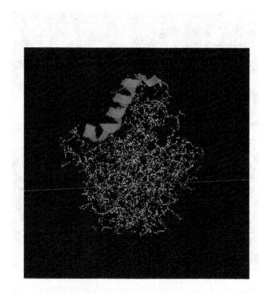

Fig. 3. Protein 1emd has an helix with width larger than average

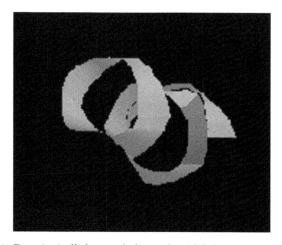

Fig. 4. Protein 1adb has an helix with width larger than average

Fig. 5. A β-strand of protein 2bpa with a width value larger than average

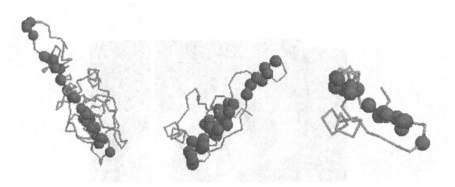

Fig. 6. (left)protein 2pf2: $\epsilon = 1$ (24 % of points,width= 18); (middle) protein 2bop: $\epsilon = 2$ (41 % of points, width=16); (right) protein 1cbn: $\epsilon = 1$ (36 % of points, width= 12)

7 Acknowledgements

Support for Guerra was provided in part by the Italian Ministry of University and Research under the National Project "Bioinformatics and Genomics Research", and by the Research Program of the University of Padova.
We would like to thank Ivano Tagliapietra for his help with this paper.

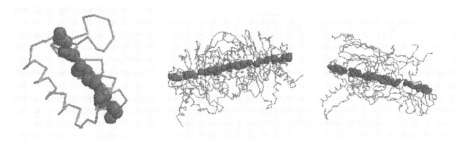

Fig. 7. (left) protein 1acp: $\epsilon = 1$ (23 % of points,width= 18); (middle) protein 4mdh: $\epsilon = 1$ (10 % of points,width= 46); (right) protein 2bpa: $\epsilon = 2$ (17 % of points,width= 57)

References

1. F.C. Bernstein, T.F. Koetzle, G.J. Williams, E.F. Meyer, M.D. Brice, J.R. Rodgers, O. Kennard, T. Shimanouchi, M. Tasumi. The Protein data bank: a computer-based archival file for macromolecular structures. *J. Mol. Biol.*,vol. 112, pp. 535-542, 1977.
2. N. Colloc'h, C. Etchebest, E. Thoreau, B. Henrissat, J.P. Mornon. Comparison of three algorithms for the assignment of secondary structures in proteins: The advantage of a consensus assignment. *Protein Eng.* , vol. 6, pp. 377-382, 1993.
3. R.O. Duda, P.E. Hart. Use of the Hough transform to detect lines and curves in pictures. *Commun. ACM,* , vol. 15, pp.11-15, 1972.
4. C.A. Duncan, M.T. Goodrich, and E.A. Ramos, Efficient Approximation and Optimization Algorithms for Computational Metrology, *8th ACM-SIAM Symp. on Discrete Algorithms (SODA)*, 1997.
5. H. Edelsbrunner. *Algorithms in Combinatorial Geometry* . Springer-Verlag, Heidelberg, Germany, 1987.
6. M. A. Fischler, R.C. Bolles. Random sample consensus: A paradigm for model fitting with applications to image analysis and automated cartography. *Computer Vision, Graphics and Image Processing*, vol. 24, pp. 381-395, n.6, 1981.
7. D. Frishman, P. Argos, Knowledge-based secondary structure assignment, *Proteins: structure, function and genetics*, vol. 23, pp. 566-579, 1995.
8. P.V. Hough. 1962. Methods and means to recognize complex patterns. U. S. patent 3.069.654.
9. M.E. Houle, G.T. Toussaint, Computing the width of a set, *IEEE Trans. on Pattern Analysis and Machine Intelligence*, PAMI-10, 1988, pp. 761-765.
10. W. Kabsch, C. Sander, Dictionary of protein secondary structure: pattern recognition of hydrogen-bonded and geometrical features, *Biopolymers*, vol. 22, pp. 2577-2637, 1983.
11. N. Kiryati, Y., Eldar, A.M., Bruckstein, A Probabilistic Hough Transform, *Pattern Recognition*, vol. 24, pp. 303-316, 1991.
12. P. Kultanen, L. Xu,E. Oja, Randomized Hough Transform (Rht), *Int. C. Pattern Recognition*, R-B(90), pp. 631-635, 1990.
13. J. Illingworth, J. Kittler, A survey of the Hough transform. *Computer Vision, Graphics and Image Processing*, vol. 44, pp. 87-116, 1988

14. M. Levitt, J. Greer. Automatic identification of secondary structure in globular protein. *J. Mol. Biol.* , vol. 114, pp. 181-239, 1977.

15. H. Li, M.A. Lavin, R.J. Le Master, Fast Hough transform. *Computer Vision, Graphics and Image Processing*, vol 36, pp. 139-161, 1986.

16. V. F. Leavers, Which Hough Transform?, *Image Understanding*, vol. 58, pp. 250-264, 1993.

17. J. Princen, J. Illingworth, J. Kittler, A Hierarchical Approach To Line Extraction Based On The Hough Transform, *Computer Vision, Graphics and Image Processing:Image Understanding*, vol. 52, pp. 57-77, 1990.

18. G. Roth, M. D. Levine, Extracting geometric primitives, *Computer Vision, Graphics and Image Processing:Image Understanding*, vol. 58, n. 1, pp. 1-22, 1993

19. L. Xu, E. Oja, Randomized Hough Transform (Rht): Basic Mechanisms, Algorithms, And Computational Complexities, *Image Understanding* vol. 57, pp. 131-154, 1993.

OPTIMA: A New Score Function for the Detection of Remote Homologs

Maricel Kann and Richard A. Goldstein*

Department of Chemistry, University of Michigan
Ann Arbor, MI 48109-1055 *kann@ncbi.nlm.nih.gov*

1 Abstract

A new method to derive a score function to detect remote relationships between protein sequences has been developed. The new score function, OPTIMA, was obtained after maximization of a function of merit representing a measure of success in recognizing homologs of the newly sequenced protein among thousands of non-homolog sequences in the databases. We find that the new score function obtained in such a manner performs better than standard score functions for the identification of distant homologies.

2 Introduction

The amount of sequence data from the rapidly progressing genome projects keeps growing exponentially. The similar fast growing speed of computers has make computer aided tools to infer structure and function of the newly sequenced proteins an integral part of modern molecular biology. In order to understand the evolutionary history of the new sequences, aligning the primary structure of the probe sequence with others in the database using sophisticated software such as FASTA [1], BLAST [2, 3], and SSEARCH [1, 4] is one the most significant and widely used techniques. Sequences with a high similarity score usually share a common structure and might have similar functions or mechanisms. For any pair of proteins, the optimal alignment that maximizes the total score can be done quickly, using standard dynamic programming techniques [5]. The maximum possible score for a given pair

---

* RAG also with the Biophysics Research Division. phone: (734) 763-8013, fax: (734)764-3323, email:richardgumich.edu.

C. Guerra, S. Istrail (Eds.): Protein Structure Analysis and Design, LNBI 2666, pp. 99-108, 2003.
© Springer-Verlag Berlin Heidelberg 2003

of proteins is then used to determine whether the pair of proteins are homologous. This is often done by computing such quantities as $p(S_r > x)$, the probability that a random pair of proteins of the same length would have that score or higher, E, the expected number of random proteins in the database that would have at least that score, and P, the probability that there is at least one random pair with a higher score. Smaller values of $p(S_r > x)$, E, and P indicate a higher likelihood that the given pair is in fact homologous.

The first commonly-used score matrices were the PAM (percent accepted mutations) series developed by Dayhoff and co-workers [6]. Others such as those developed by Gonnet et al (GCB) [7] and Jones et al. (JTT) [8] have applied Dayhoff's method to larger sequence datasets. Henikoff and Henikoff used a dataset of aligned sequence blocks to construct his popular BLOSUM62 matrix [9]. Overington and coworkers used Henikoffs' cluster method to create a score matrix (STR) where the protein sequences were aligned based on their observed structures [10].

The matrices described above, represent considerable improvements in the search procedure, allowing the immediately detection of similarities among approximately half of all the newly discovered genes. However, there is still a challenging problem to overcome, when sequence similarity is low with those score functions (below 25% sequence identity) distant relationships among protein sequences that are in fact homologs, are impossible to detect.

In this paper, we describe a new procedure to generate a score function optimized to detect distantly related pairs of protein sequences. Results with the independent test sets demonstrate the superiority of the resulting score function compared with other commonly-used score functions for the detection of distant homologies.

3 Methods

Theory

We sequentially align a target protein A_t with each of the proteins in a dataset of size D, achieving a distribution of scores $\{S_r\}$ of the form

$$S = \sum_{i,j} n_{i,j}\, \gamma_{i,j} + n_{\text{gap}-\text{I}}\, \gamma_{\text{gap}-\text{I}} + n_{\text{gap}-\text{E}}\, \gamma_{\text{gap}-\text{E}} \tag{1}$$

where $n_{i,j}$ refers to the number of times that amino acid type i is aligned with amino acid type j, $n_{\text{gap}-\text{I}}$ is the total number of gaps in the alignment, $n_{\text{gap}-\text{E}}$ is the total number of residues in each gap beyond one, $\gamma_{i,j}$, also known as the score function, substitution matrix, or exchange residue matrix, represents the contribution to the score for any amino acid match or mismatch, while the gap penalties are given by $\gamma_{\text{gap}-\text{I}}$ and $\gamma_{\text{gap}-\text{E}}$ initialization and extention of a gap, respectively . The score for the alignment of the target protein and a putative homolog is x. To estimate the significance of this score, we calculated the probability that this score or higher would be obtained by a random match. We first compute the Z-score, defined as

$$Z = \frac{x - \langle S_r \rangle}{\sqrt{\langle S_r^2 \rangle - \langle S_r \rangle^2}} \qquad (2)$$

where the averages are over the alignments of the target protein with the ensemble of random non-homologous proteins in the dataset. By using the Z score, we automatically account for variations in the expected score with the length of the proteins. We can represent the distribution of scores for ungapped [3, 14, 15] and gapped alignments [16] by an extreme value distribution (EVD) [17]. In this case, the probability that a given random score S_r would be equal or greater than x is

$$p(S_r > x) = 1 - \int_{-\infty}^{S_r} \rho(x) \, dx$$
$$= 1 - \exp\left(-\exp\left(-\alpha Z - \beta\right)\right) \qquad (3)$$

$\alpha = 1.282$ and $\beta = 0.5772...$ for a perfect EVD [18], although these parameters are generally adjusted based on the observed distribution. For a search of a database of size D, the expected number of scores between the target protein and random pairs is equal to $E = D \, p(S_r > x)$. In this paper, we use a value of $D = 100,000$. Assuming a Poisson distribution, the probability P of observing at least one alignment with score equal to or greater than x is given by

$$P = 1 - \exp\left(-E\right) \qquad (4)$$

Both the E-value and the P-value depend upon the size of the database being searched. E-values range from 0 to D, while P-values range from 0 to 1.

We are interested in optimizing the ability of a score function to discriminate between homologous and non-homologous pairs of proteins. That is, we are interested in identifying a true homolog, and in having confidence in this identification. Our confidence in a putative match is equal to the number of correct matches divided by the number of matches, both correct and incorrect, with the same score or higher. Assuming that we have one true homolog in the dataset, the average confidence C can be quantified as

$$C = \frac{1}{1 + E} = \left(1 + D\left(1 - e^{-\exp(-\alpha Z - \beta)}\right)\right)^{-1} \qquad (5)$$

A C value close to 1 indicates a confident alignment, with C decreasing to $1/(1 + D)$ as the confidence of the alignment decreases. This represents our average relative chance the match is to a true homolog. In this paper we optimize the score function by maximizing $\langle C \rangle$ averaged over the training set. By optimizing $\langle C \rangle$ we automatically focus on homologous pairs at the limit of detection, reducing the dependence of the score function on homologies that are either easily detectable ($E \ll 1$) or overly distant ($E \sim D$).

Database Preparation

We are interested in optimizing our score functions for the detection of distant homologs, beyond the capability of current score functions. We therefore need a set of known homologs whose homology cannot be reliably determined with standard pairwise sequence comparisons. For this purpose, we took advantage of the Cluster of Orthologs Genomes (COG) database of Koonin and co-workers [13]. A 900-pair training set was constructed of pairs of proteins in the same COG that share less than 25% sequence identity. A 177-pair disjoint test set was constructed in a similar manner from the COG database, excluding all COGs that contributed to the training set, with each pair of proteins again having less than 25% sequence identity. We also constructed a test sets independently of the method used to construct the training set by taking pairs of proteins identified as homologs in the PFAM database release 5.2 [19]   In order to avoid overlap with the training and test sets derived from the COG database, we ran a BLAST search [2] (using BLOSUM62 [9] with -12,-2 for the gap penalties) of all the sequences in the PFAM database against the 1077 pairs from the COGs that we were using either as the training set or first test set, and excluded all PFAM families with any member with similarity to these proteins (i.e. $E < 10$). From this set of protein sequences, we selected 103 pairs that share less than 25% sequence identity as a second test set of distantly related sequences.

Optimization of the Score Function

We are interested in maximizing the confidence value $\langle C \rangle$ averaged over proteins in the training set, where the calculation of C involves the distribution of scores for the optimal alignment of the target proteins with homologous and non-homologous proteins in the dataset. These optimal alignments are themselves dependent upon the value of the score function. Thus an iterative scheme is required.

   We started with the BLOSUM62 matrix [9] and used the local dynamic programming algorithm [5] to align each of the target proteins in the training dataset against a homolog and a set of non-homologous proteins with a large number of different gap penalties.   We then calculated Z and C for each pair of homologs, and averaged over the pairs in the training set to yield $\langle C \rangle$. The highest C values were obtained with gap penalties of $\gamma_{\text{gap}-\text{I}} = -12.0$ and $\gamma_{\text{gap}-\text{E}} = -2.0$. This scoring scheme (BL62(12,2)) was then used to generate an initial set of alignments of the target proteins with homologous and non-homologous proteins. The observed distributions of the non-homologous proteins were used to adjust the values of $\alpha = 1.31$ and $\beta = 0.74$, similar to the values expected ($\alpha = 1.282$ and $\beta = 0.5772$) for a perfect EVD [18].

   As multiplication of the score function by any constant does not change Z or any of the other statistics, we fixed one score function ($\gamma_{\text{gap}-\text{I}}$) equal to

-2.0, resulting in 211 adjustable parameters corresponding to the 210 possible pairs of amino acid types and the remaining gap penalty. Starting with the BL62(12,2) score scheme and the corresponding set of aligned sequences, we analytically calculated the approximate direction of a steepest descent for the adjustable parameters assuming the alignments remained unchanged. We then adjusted the parameters along that direction, realigning the sets of sequences at every point, until Armijo's rule was satisfied [20]. The next appropriate direction of steepest descent was then recalculated. Approximately 10 cycles of optimization and re-alignments were performed until the score function converged. Performance was monitored by simultaneous calculations of $\langle C \rangle$ averaged over the proteins in the test set of distant homologs derived from the COG database. The statistics with the optimized score function indicated that the appropriate values of α and β did not appreciably shift.

Results

The values of $\langle C \rangle$ as averaged over the training set and distant COG homolog test set during the optimization process are shown in Figure 1. The resulting score function (OPTIMA) obtained after 10 iterations is shown in Table I.

Figure 2 shows the cumulative distribution of C values for COG distant-homolog test set with the different score functions. As shown, the greater discriminatory power of the OPTIMA score function is represented by the larger fraction of the protein sequence pairs having larger values of C. That implies a substantial improvement in our ability of making confident predictions compared with other standard score functions. The optimal value for

	A	R	N	D	C	Q	E	G	H	I	L	K	M	F	P	S	T	W	Y	V
A	36																			
R	-9	56																		
N	-19	4	59																	
D	-20	-18	18	65																
C	6	-29	-30	-30	99															
Q	-3	12	2	2	-30	46														
E	-10	3	3	20	-39	19	40													
G	4	-18	7	-10	-29	-20	-23	67												
H	-19	3	12	-7	-29	3	2	-18	86											
I	-5	-28	-32	-34	-6	-30	-33	-41	-28	35										
L	-7	-20	-32	-43	-6	-23	-31	-42	-27	28	32									
K	-10	31	1	-4	-29	15	14	-18	-7	-31	-21	37								
M	-9	-10	-19	-30	-8	1	-21	-30	-19	12	24	-12	51							
F	-19	-30	-29	-33	-18	-29	-32	-32	-8	8	17	-29	2	57						
P	-5	-18	-17	-7	-30	-11	-7	-18	-18	-30	-33	-10	-21	-39	74					
S	12	-11	10	4	-10	0	-1	2	-9	-20	-22	3	-10	-19	-8	36				
T	0	-8	0	-10	-7	-7	-6	-17	-20	-8	-13	-8	-7	-18	-11	18	48			
W	-29	-29	-39	-40	-18	-19	-29	-19	-18	-28	-15	-30	-8	14	-38	-29	-19	110		
Y	-19	-15	-19	-20	-18	-9	-21	-29	20	-8	-2	-17	-9	37	-28	-19	-17	22	69	
V	6	-32	-31	-31	-6	-19	-28	-30	-29	35	18	-23	10	0	-18	-22	6	-28	-9	38

Table 1. OPTIMA score matrix achieved at the tenth iteration. The elements are multiplied by ten to increase precision; corresponding gap penalties are -120 and -20.

$\gamma_{\text{gap}-\text{I}}$ was -11.97 with $\gamma_{\text{gap}-\text{E}}$ fixed at -2.0. The small change in the gap penalties indicates that most of the improvement comes from refinements of

	Score Matrix	Gap penalties	COGs Distant Homologs			PFAM Distant Homologs		
		(Initiate/Extend)	$\langle C \rangle$	$\langle p(S_r > x) \rangle$	$\langle P \rangle$	$\langle C \rangle$	$\langle p(S_r > x) \rangle$	$\langle P \rangle$
	OPTIMA	-11.97/-2.0	0.736	0.004	0.283	0.710	0.007	0.311
1	BLOSUM62 [9]	-12/-2	0.652	0.009	0.372	0.645	0.016	0.376
2	BLOSUM62 [9]	-8/-0.5 [22]	0.248	0.024	0.800	0.312	0.033	0.737
3	PAM250 [6]	-12/-2	0.480	0.017	0.549	0.669	0.012	0.359
4	PAM250 [6]	-6/-1.3 [22]	0.013	0.072	0.999	0.035	0.061	0.988
5	GCB [7]	-12/-2	0.647	0.007	0.377	0.703	0.022	0.324
6	GCB [7]	-7.5/-0.4 [22]	0.023	0.061	0.997	0.030	0.042	0.987
7	STR [10]	-12/-2	0.515	0.035	0.509	0.575	0.033	0.450
8	STR [10]	-8/-0.5 [22]	0.172	0.041	0.866	0.281	0.034	0.774
9	JTT [8]	-12/-2	0.517	0.009	0.516	0.642	0.022	0.392
10	JTT [8]	-10.5/-0.4 [22]	0.076	0.035	0.958	0.127	0.036	0.916

Table 2. Comparison of the various score matrices and gap penalties on PFAM and COG test sets of distant homologs (percentage identity less than 25%) as evaluated with average confidence value ($\langle C \rangle$), average probability that a random score would be higher than the known homolog ($\langle p(S_r > x) \rangle$), and average probability that at least one of the random scores in a dataset of 100,000 proteins would be higher than the known homolog ($\langle P \rangle$). For the purpose of these comparisons, we use both the standard default gap penalties as well as the gap penalties derived by Argos [22].

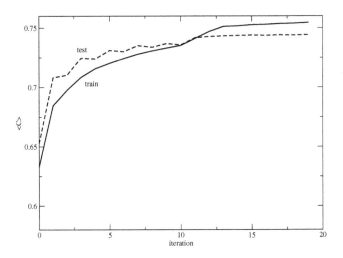

Fig. 1. Value of $\langle C \rangle$ averaged over the training dataset and test dataset of distant homologs during the optimization procedure.

the values of $\gamma_{i,j}$. As shown in Table II, OPTIMA has a significantly improved average confidence ($\langle C \rangle$) value compared with other commonly-used score matrices . This improvement is not confined only to values of C; both $\langle p(S_r > x) \rangle$, the average probability that any random score would be higher than the homolog, as well as $\langle P \rangle$, the average probability that at least one random score is higher than the homolog, are both substantially decreased compared with other matrices.

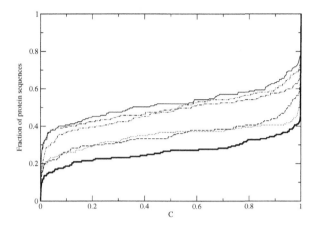

Fig. 2. Cumulative distribution of the C values for the various score matrices, showing the fraction of all protein pairs in the COGs test set of distant homologs with less than a given value of confidence. The various lines refer to the OPTIMA score matrix (–); BLOSUM62 (12/2)[9] ($\cdots\cdots$); GCB (12/2)[7] (– – –); STR (12/2)[10] (– · – · – · –); JTT (12/2)[8] (– · · – · · –); and PAM250 (12/2)[6] (–).

Figures 3 A and B show the coverage or fraction of true positives vs. the estimated number of false positives per query for the distant homolog COGs and PFAM test sets, respectively, where the estimated number of false positives per query represents the expected number of random sequences with a score greater than the pair of homologous sequences. The better performance of OPTIMA can be seen from the large number of homologous pairs with lower estimated number of false positives.

As a further test, we constructed a parametric plot where we calculated the fraction of true positive homologs identified (coverage) and the fraction of non-homologous pairs identified incorrectly as homologies (fraction of false positives) as a function of the cut-off value of E. The results are shown for the distant homolog COGs and PFAM test sets in Figures 3 C and D, respectively. While this parametric plot may be compromised by the presence of true homologs incorrectly labeled as non-homologs (false false-positives) in the test sets, the qualitative agreement with the previous plots further supports the performance of OPTIMA compared with the other score functions.

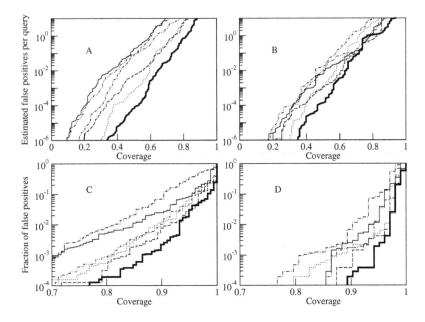

Fig. 3. A) Expected number of false positives of different score matrices as a function of the number for protein sequences pairs (Coverage) for the COG test dataset of distant homologs. The plots show the fraction of the homologous pairs with fewer than a given number of false positives of higher score (E) expected for $D = 100,000$. **B)** Similar plot for the PFAM test dataset of distant homologs. **C)** Coverge vs. number of false positives for the COG test dataset of distant homologs. For this plot, we calculated the E values for each homologous and non-homologous pair of proteins in the test set, and then ranked these values from lower to higher. We then considered all matches with a value of E lower than a given cut-off value, and calculated the fraction of the true positives included in this set (Coverage) as well as the fraction of non-homologous pairs also included (Fraction of false positives). These two values were then plotted as a parametric plot. This approach, applied to homology detection by Brenner et al. [11], is related to the Receiver Operating Characteristic (ROC) measure [23, 24]. **D)** Similar plot for the PFAM test dataset of distant homologs. All of the curves are designated as in Figure 2

Discussion

Most methods for constructing a score function rely on creating a dataset of reliably-aligned sequences or sequence fragments and gathering statistics on the relative number of times that each possible pair of amino acids are aligned. In practice, however, we are interested in distinguishing optimally (but possibly incorrectly) aligned homologs from optimal alignments of non-homologs. Our approach towards generating a score matrix is to optimize the ability of this matrix to do what we want to do: discriminate between homologs and non-homologs. In order to do this we derived a measure of merit of the score function, the average confidence of the homolog identification, and maximized this measure over a set of homologous and non-homologous pairs of proteins. Different measures of merit can be handled in a similar way. Our score function still represents statistics derived from real homologous protein sequences, and can be analyzed in terms of evolutionary substitutions and the physical-chemical properties of the amino acids. In contrast to standard derivations, the gap penalties can be treated as additional parameters to be optimized. In tests with two disjoint set of test proteins, we are able to demonstrate that this score function achieves greater success at discriminating between homologs and non-homologs compared with standard score matrices.

Acknowledgements. The authors would like to thank Bin Qian for his collaboration to this project and Stephen Altschul, Ting-Lan Chiu, William Pearson, Romesh Saigal, Stephen Bryant and John Spouge for helpful discussions, also Todd Raeker and Michael Kitson for computational assistance. Financial support was provided by NIH Grant LM0577 and NSF equipment grant BIR9512955.

References

1. Pearson, W. R., Lipman, D. J. Improved tools for biological sequence analysis. Proc. Nat. Acad. Sci. USA 85:2444–2448, 1988.
2. Altschul, S. F., Gish, W., W.Miller, Myers, E., Lipman, D. Basic local alignment tool. J. Mol. Biol. 215:403–410, 1990.
3. Altschul, S. F., Gish, W. Local alignment statistics. Methods Enzymol. 215:460–480, 1996.
4. Lipman, D. J., Pearson, W. R. Rapid and sensitive protein similarity searches. Science 227:1435–1441, 1985.
5. Smith, T. F., Waterman, M. S. Identification of common molecular subsequences. J. Mol. Biol. 147:195–197, 1981.
6. Dayhoff, M. O., Schwartz, R. M., Orcutt, B. C. A model of evolutionary change in proteins. In: Atlas of Protein Sequence and Structure. Vol. 5. Dayhoff, M. O. ed. Vol. 5. . National Biomedical Research Foundation 1978.
7. Gonnet, G. H., Cohen, M. A., Benner, S. A. Exhaustive matching of the entire protein database. Science 256:1443–1445, 1992.
8. Jones, D. T., Taylor, W. R., Thornton, J. M. The rapid generation of mutation data matrices from protein sequences. CABIOS 8:275–282, 1992.
9. Henikoff, S., Henikoff, J. G. Amino acid substitution matrices from protein blocks. Proc. Nat. Acad. Sci. USA 89:10915–10919, 1992.
10. Overington, J., Donnelly, D., Johnson, M. S., Šali, A., Blundell, T. L. Environment-specific amino-acid substitution tables: Tertiary templates and prediction of protein folds. Protein Sci. 1:216–226, 1992.
11. Brenner, S. E., Chothia, C., Hubbard, T. J. P. Assessing sequence comparison methods with reliable structurally identified distant evolutionary relationships. Proc. Nat. Acad. Sci. USA 95:6073–6078, 1998.

12. Altschul, S. F., Madden, T. L., Schaffer, A. A., Zhang, J., Zhang, Z., Miller, W., Lipman, D. L. Gapped Blast and Psi-Blast: A new generation of protein database search programs. Nucl. Acid Res. 25:3389–3402, 1997.

13. Tatusov, R. L., Galperin, M. Y., Koonin, E. V. The COG database: a tool for genome-scale analysis of proteins functions and evolution. Nucleic Acids Res 28:33–36, 2000.

14. Karlin, S., Altschul, S. F. Methods for assessing the statistical significance of molecular sequence features by using general scoring schemes. Proc. Nat. Acad. Sci. USA 87:2264–2268, 1990.

15. Dembo, A., Karlin, S., Zeitouni, O. Limit distribution of maximal non-aligned two-sequence segmental score. Ann. Prob. 22:2022, 1994.

16. Pearson, W. R. Empirical statistical estimates for sequence similarity searches. J. Mol. Biol 276:71–84, 1998.

17. Gumbel, E. J. Statistics of Extremes. New York: Columbia University Press. 1958.

18. Gumbel, E. J. Statistics theory of extreme values and some practical applications. Washington: U.S. Government Printing Office: National Bureau of Standards Applied Mathematics Series 33. 1954.

19. Bateman, A., Birney, E., Durbin, R., Eddy, S. R., Finn, R. D., Sonnhammer, E. L. L. Pfam 3.1: 1313 multiple alignments match the majority of proteins. Nucleic Acids Research 27:260–262, 1999.

20. Dennis, J. E., Jr., and Schnabel, R. B. Numerical Methods for Unconstrained Optimization and Nonlinear Equations. New York: Prentice-Hall. 1983.

21. Press, W. H., Teukolsky, S. A., Vetterling, W. T., Flannery, B. P. Numerical Recipes in C: Cambridge University Press. 1992.

22. Vogt, G., Etzold, T., Argos, P. An assessment of amino acid exchange matrices in aligning protein sequences: The twilight zone revisited. J. Mol. Biol. 249:816–831, 1995.

23. Metz, C. E. ROC methodology in radiologic imaging. Invest. Radiol. 21:720–733, 1986.

24. Swets, J. A. Measuring the accuracy of diagnostic systems. Science 240:1285–1293, 1988.

A Comparison of Methods for Assessing the Structural Similarity of Proteins

Dean C. Adams and Gavin J.P. Naylor*

Dept. Zoology and Genetics, Iowa State University, Ames, IA 50011, U.S.A.
{dcadams, gnaylor}@iastate.edu

1 Introduction

The link between biological form and function is well known, and is assumed to hold true at the molecular level. Since identifying similar protein structures is the first step in identifying similar functions, much effort has been placed in developing methods to detect structural similarity. Several methods exist, including: SCOP [8], the DALI algorithm (from the FSSP Database [6]), the VAST algorithm (from the MMDB database [5]), and Root Mean Square (RMS) superimposition [9]. The latter three provide quantitative metrics describing protein similarity on an objective, continuous scale. Statistical analyses can then be performed on similarity scores for a set of proteins, to obtain a plot of 'protein structure space' [7]. Before such analyses are done however, one must be sure that the metric used accurately represents similarity.

In this paper, we describe the DALI Z-score and RMS-distance (D_{RMS}) metrics, and discuss their shortcomings. We then present a novel means of comparing protein structures using Geometric Morphometric (GM) methods: statistical shape methods borrowed from anatomy. Finally, we compare results from these three methods for a data set of globin structures, and show that the more intuitive GM method markedly outperforms existing techniques.

2 The DALI Algorithm

The DALI algorithm [6],[7] compares protein structures using two-dimensional matrices, where each element in the matrix (d_{ij}) is the Euclidean distance

---

* Work supported in part by a National Science Foundation Postdoctoral Fellowship in Biological Informatics: DBI-9974207 to DCA.

C. Guerra, S. Istrail (Eds.): Protein Structure Analysis and Design, LNBI 2666, pp. 109-115, 2003.

between the ith and j^{th} residues for that protein. Distance matrices are aligned in pairwise fashion, and n homologous residues are identified. The structural similarity for the two proteins (A and B) is then calculated as:

$$S = \sum_i \sum_j \left(0.2 - \frac{|d_{ij}^A - d_{ij}^B|}{d_{ij}^*} \right) e^{-\left(\frac{d_{ij}^*}{20\text{Å}}\right)^2} \tag{1}$$

where d_{ij}^* is the mean distance for those residues (a standardized version of S, the Z-score, is also calculated). Z-scores are calculated for all protein pairs, and the best three-dimensional ordination of the structure space is found through an eigen-decomposition (correspondence analysis) of the Z-score matrix, where similar proteins are close together, and dissimilar proteins are far apart.

Though Z-scores quantify some aspects of structural similarity, details of this metric warrant careful scrutiny. First, Z-scores are generated from pairwise alignments, so different residues can be used for each pair. Thus, values in the Z-score matrix represent different aspects of structural similarity, and are not directly comparable. Second, the metric contains a dissimilarity cut-off (0.2) to eliminate protein comparisons > 20%. However, most protein comparisons in a large database are > 20%, yielding negative scores, which are arbitrarily truncated to zero. An eigen-analysis of such data will explain little of the variation in few dimensions, and a low dimensional ordination from this analysis will fail to capture the essence of 'structure space.' Finally, the exponential term in the metric downweights contributions from residues far from one another. This results in Z-scores for self-comparisons that are not the same for each protein, implying that some proteins are more 'perfectly' similar to themselves than others, which is nonsensical. Thus, DALI Z-scores are not a true similarity metric, and statistical analyses of them are unpredictable.

3 The Root Mean Square Algorithm

Root Mean Square (RMS) methods assess structural similarity using a least squares (LS) criterion. First, two proteins (X & Y) are structurally aligned to identify the set of n homologous residues [4]. Next, they are translated to a common location, and are rotated so that homologous residues line up as closely as possible [9]. Finally, the Euclidean distance (D_{RMS}) between them is calculated:

$$D_{RMS} = \sqrt{\sum_{i=1}^{n} \sum_{j=1}^{3} (X_{ij} - Y_{ij})^2} \tag{2}$$

where X_{ij} and Y_{ij} are the coordinate sets for the ith residue. D_{RMS} is calculated for all protein pairs, and the best three-dimensional ordination of the structure space is found through an eigen-decomposition (principal coordinates analysis) of the D_{RMS} matrix.

RMS methods are appealing because D_{RMS} makes intuitive sense: unlike proteins have a large D_{RMS}, while similar proteins align quite well and have a small D_{RMS}. It is also a true distance measure, because all self-comparisons of proteins yield an identical value of zero (no structural differences). Like Z-scores however, D_{RMS} is calculated in pairwise fashion, so different residues can be used for each protein pair, rendering D_{RMS} scores incomparable.

4 Geometric Morphometrics

Both DALI Z-scores and D_{RMS} can be used to generate a map of protein structure space. However, both have methodological difficulties which limit their utility. Interestingly, these same difficulties have already been addressed in a completely different discipline: Geometric Morphometrics (GM). GM methods were originally developed to analyze anatomical structures (e.g., skulls), but may easily be adapted to compare macromolecular structures. First, a set of homologous points recorded on all specimens are superimposed using generalized Procrustes analysis (GPA), which translates specimens to a common location, scales them to unit size, and optimally rotates them (in a LS sense) [3],[10]. Shape variables are then generated for each specimen, which may be used in statistical analyses [2]. Additionally, Procrustes distance (D_{PROC}) between two specimens (X & Y) can be calculated as:

$$D_{PROC} = 2\sin^{-1}\left(\sqrt{\sum_{i=1}^{n}\sum_{j=1}^{3}(X_{ij} - Y_{ij})^2/2}\right) \tag{3}$$

where X_{ij} and Y_{ij} are the aligned coordinates for the ith residue. D_{PROC} is calculated for all protein pairs, and the best three-dimensional plot of protein shape space is found through an eigen-decomposition (principal coordinates analysis) of this data. Although GM and RMS are quite similar, they differ in two important respects. First, size is mathematically held constant in GPA (not in RMS), and second, GPA superimposes all specimens simultaneously.

5 Comparison of Methods

To compare the three methods described above we used a representative set of protein structures. We extracted all globin sequences (as of 12/10/1999) from the Protein Data Bank, and separated them into their individual chains, so that monomeric and non-monomeric globins could be used. Structural similarity among the 560 chains was then assessed using each of the three methods (Z-scores, D_{RMS}, D_{PROC}). Pairwise structural alignments were calculated in the DALI domain dictionary [7] (http://www2.ebi.ac.uk/dali) and both DALI Z-scores and D_{RMS} scores were obtained for each protein pair. For GM, we aligned the amino acid sequences with Clustal W [11] and deleted all gaps, yielding 96 homologous residues [1]. We then superimposed the structural data for these residues with GPA, and generated D_{PROC} for each protein pair.

The ability of each metric to capture structural variation was assessed using multivariate ordination methods. The DALI Z-score matrix was summarized using correspondence analysis (as per [7]), and the D_{RMS} and D_{PROC} matrices were summarized using principal coordinates analysis. The percentage of variation explained by the first three dimensions from the ordination analysis was compared for each method, and their ability to identify biologically meaningful clusters was assessed through a visual inspection of the ordination plots.

Using D_{RMS}, the 1st three dimensions of structure space explained 76.1% of globin chain variability. Inspection of this ordination plot revealed separation of a few individual chains (mostly hemoglobin chains), but no obvious groups were identifiable (Fig. 1). Thus, although D_{RMS} explained much of the variation, it was unable to identify any biologically meaningful globin clusters.

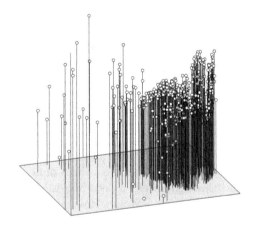

Fig. 1. Three-dimensional view of globin structure space from D_{RMS}.

Using DALI Z-scores, the 1st three dimensions of structure space explained 33.5% of globin chain variation, and to describe an equivalent amount of variation to D_{RMS} (76%), 56 dimensions of the ordination were needed. Further, the ordination plot revealed no obvious clusters of globin chains (Fig. 2). Thus, DALI Z-scores were much less effective at summarizing structural variability, and were unable to reveal biologically interpretable clusters of proteins.

Using D_{PROC}, the 1st three dimensions of the GM shape space explained 76.6% of the variation, which was similar to that found with D_{RMS}. However, the ordination plot revealed remarkable separation of globin chains into

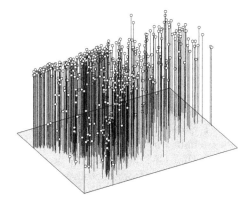

Fig. 2. Three-dimensional view of globin structure space from DALI Z-scores.

identifiable groups. These groups corresponded to meaningful biological partitions of the data set, including: bacterial hemoglobins, clam hemoglobins, ferric hemoglobins, hemoglobin α (& β)-chains, lamprey hemoglobins, leghemoglobins, and myoglobins (Fig. 3). Thus, much more biological information is obtained using GM, as compared to either DALI Z-scores or D_{RMS}.

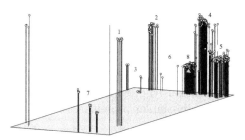

Fig. 3. Three-dimensional view of globin shape space from D_{PROC}. Labels correspond to the following groups: 1: bacterial hemoglobins, 2: clam hemoglobins, 3: ferric globins, 4: hemoglobin α-chains, 5: hemoglobin β-chains, 6: lamprey hemoglobins, 7: leghemoglobins, and 8: myoglobins.

6 Discussion

In this paper we described two metrics used for assessing structural similarity (DALI Z-scores and D_{RMS}), and described how geometric morphometric (GM) methods, commonly used in anatomical studies, may also be employed to compare protein structures. We then compared the ability of three metrics to summarize structural variation in a set of globin structures. DALI Z-scores explained very little (33.5%) of the total variation in three dimensions, and were unable to identify any globin clusters. D_{RMS} explained significantly more variation (76.1%) in three dimensions, but it too was unable to identify clusters of globins. On the other hand, D_{PROC} explained a large proportion of the variation (76.6%) in three dimensions, *and* was able to identify biological clusters of globins (e.g., bacterial hemoglobins, leghemoglobins, myoglobins, etc.). Further, all but 1 globin chain was correctly classified to its biological group (1 hemoglobin α-chain was classified as hemoglobin β-chain).

These results suggest that GM methods may be more useful for extracting meaningful biological information from protein structures than are either the DALI or RMS methods. Why might this be the case? It seems that DALI Z-scores are predisposed *not* to identify meaningful structural variation: they are calculated in pairwise fashion, and are not a true similarity measure. Further, the arbitrary similarity cut-off predisposes them to identify many protein comparisons as 'maximally' different. Explaining the performance of D_{RMS} however, is more difficult. D_{RMS} is a true distance measure, so it does not suffer the same problem as Z-scores. Further, D_{RMS} and D_{PROC} differ algorithmically in only two respects: D_{RMS} is a linear distance (where D_{PROC} is curve-linear), and the D_{PROC} protocol standardizes the size of each specimen (but for globins, size accounts for a tiny portion of variation). The only other difference between the RMS and GM methods is that D_{RMS} is calculated from homologous residues from *pairwise* structural alignments, whereas D_{PROC} is calculated from homologous residues common to *all* structures (found from a multiple alignment). This assures that the information contained in D_{PROC} is consistent among comparisons, which is an explicit requirement of any statistical analysis. It appears then, that the GM protocol, using homology defined for *all* proteins simulataneously, provides the best chance for identifying natural clusters of structurally-similar proteins.

References

1. D. C. Adams and G. J. P. Naylor: A new method for evaluating the structural similarity of proteins using geometric morphometrics. in S. Miyano, R. Shamir, and T. Takagi (eds.) *(*Currents in computational molecular biology). Universal Academy Press, Tokyo 2000.
2. D. C. Adams and F. J. Rohlf: Ecological character displacement in *Plethodon*: biomechanical differences found from a geometric morphometric study. Proc. Natl. Acad. Sci. U.S.A. **97** (2000) 4106–4111.
3. F. L. Bookstein: *Morphometric Tools for Landmark Data: Geometry and Biology*. Cambridge University Press, Cambridge 1991.
4. M. Gerstain and M. Levitt: Comprehensive assessment of automatic structural alignment against a manual standard, the scop classification of proteins. Protein Sci. **7** (1998) 445–456.
5. J.-F. Gibrat, T. Madej, and S. H. Bryant: Surprising similarities in structure comparison. Curr. Opin. Struct. Biol. **6** (1996) 377–385.

6. L. Holm and C. Sander: Protein structure comparison by alignment of distance matrices. J. Mol. Biol. **233** (1993) 123–138.
7. L. Holm and C. Sander: Mapping the protein universe. Science **273** (1996) 595–602.
8. A. G. Murzin, S. E. Brenner, T. Hubbard, and C. Chothia: SCOP: A structural classification of proteins database for the investigation of sequences and structures. J. Mol. Biol. **247** (1995) 536–540.
9. S. T. Rao and M. G. Rossman: Comparison of super-secondary structures in proteins. J. Mol. Biol. **76** (1973) 241–246.
10. F. J. Rohlf and D. E. Slice: Extensions of the Procrustes method for the optimal superimposition of landmarks. Syst. Zool. **39** (1990) 40–59.
11. J. D. Thompson, D. G. Higgins, and T. J. Gibson: Clustal W: Improving the sensitivity of progressive multiple sequence alignment through sequence weighting, position specific gap penalties and weight matrix choice. Nucl. Acids Res. **22** (1994) 4673–4680.

Prediction of Protein Secondary Structure at High Accuracy Using a Combination of Many Neural Networks

Claus Lundegaard[1], Thomas Nordahl Petersen[1], Morten Nielsen[1], Henrik Bohr[2], Jacob Bohr[2], Søren Brunak[2], Garry Gippert[1], and Ole Lund[1]

[1] Structural Bioinformatics Advanced Technologies A/S, Agern Allé 3, DK-2970 Hørsholm, Denmark
[2] Structural Bioinformatics Inc., SAB, San Diego, California

1 Summary

A protein secondary structure prediction protocol involving up to 800 neural network predictions has been developed by SBI-AT. An overall performance of 80% is obtained for three-state (helix, strand, coil) DSSP categories. Input to primary-layer neural networks includes sequence profiles, relative residue position, relative chain length, and amino-acid composition. Secondary structure predictions are made for three consecutive residues simultaneously – a technique which we describe as 'output expansion' – which boosts the performance of second-layer structure-to-structure networks. Independent network predictions arise from 10-fold cross validated training and testing of 1032 protein sequences at both primary and secondary network layers. Network output activities are converted to probabilities. Finally, 800 different predictions are combined using a novel balloting procedure.

2 Introduction

Prediction of three-dimensional protein structure from primary amino-acid sequence is one of the biggest challenges in structural biology today. One step toward solving this problem is by increasing the accuracy of secondary structure predictions for subsequent use as input to ab initio calculations or threading algorithms. The predictions are of significant importance to protein fold recognition and homology modeling, and the degree to which the secondary structure can be determined has become a benchmark for protein

C. Guerra, S. Istrail (Eds.): Protein Structure Analysis and Design, LNBI 2666, pp. 117-122, 2003.
© Springer-Verlag Berlin Heidelberg 2003

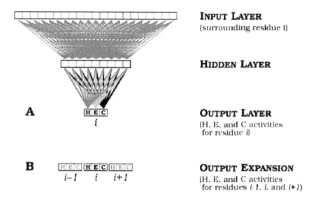

Fig. 1. Architecture of first-layer neural networks used for secondary structure predictions (A) without output expansion, (B) with output expansion.

structure prediction. The most successful method of secondary structure prediction so far has been the use of artificial neural networks. Early methods for these predictions relied on the use of a single protein sequence [5, 3]. However, a significant increase in performance can be obtained by using sequence profiles as demonstrated with the PHD method developed by Rost and Sander [12, 13]. This method performed best in the CASP2 experiment with a mean Q3 (three state prediction: helix, strand, coil) of 74%. In CASP3 the PSI-PRED method [7] performed best with a Q3 performance approximately seven percentage points better than a version of the PHD method similar to the one used in CASP2 [10]. The major difference between theese two methods is that Rost and Sander used probability scores obtained by a single Blast search [1], whereas Jones used the probability scores obtained from several iterative search rounds using PSI-Blast [2]. Studies have shown that an increased performance in secondary structure prediction can also be obtained by combining several estimators, such as neural networks [12, 6]. A combination of up to eight neural networks has been shown to increase the accuracy, but a saturation point was reached in the sense that adding more networks would not increase the performance substantially [4]. The present paper is a description of the performance of an improved procedure for combining up to 800 predictions generated using differently trained neural networks [11].

3 Methods

3.1 Neural Networks

A feed-forward neural network architecture was used, with input, hidden and output layers (Figure 1). Input consists of sequence profiles in window sizes of 15, 17, 19 and 21 residues surrounding position i, relative residue position, relative chain length and per-chain amino-acid composition. Hidden layer sizes of 50 and 75 units were used.

In conventional networks the output layer (Figure 1A) consists of sigmoidal activities for H, E and C only for position i. Using output expansion the output layer (Figure 1B) consists of HEC activities also for positions neighboring i. First-layer network output is fed into a second-layer structure-to-structure network using an input window size of 17, hidden size 40, and output HEC activities for residue i.

3.2 Data Set Used for Network Training

Network training data was prepared from the PDB version of August 1999. A set of 1168 protein chains remained after an extensive quality check [11], application of a strict pairwise sequence similarity threshold [9], and removal of transmembrane spanning chains. Eight-category DSSP secondary structure assignments [8] were reduced to three-category H (helix), E (extended strand) and C (all other) for use in network training. Sequence profiles were obtained using iterative BLAST searches [2]. The set of 1168 protein chains was further homology reduced against the evaluation set prior to network training, yielding a set of 1032 chains representing ∼200,000 amino-acid residues.

3.3 Evaluation Data Set

Prediction performance was measured using a benchmark set of 126 protein chains [12]. Several sequences in the 126 set were found to contain structural gaps (missing residues) in the corresponding PDB entry. Missing segments obtained from SWISS-PROT sequences were included as context information for residues with DSSP structural assignments.

3.4 Balloting Procedure

Balloting consists of taking the weighted average of the subset of predictions with 'confidence' greater than a specified threshold, expressed as a Z-score with respect to the mean. The subset of networks is in general different for each query chain. Per-chain, per-prediction confidence is the mean difference between highest and next-highest probability among three-category predictions [11].

4 Results

A performance of 80% mean per-residue secondary structure prediction accuracy was obtained using a benchmark set of 126 protein sequences [11, 12].

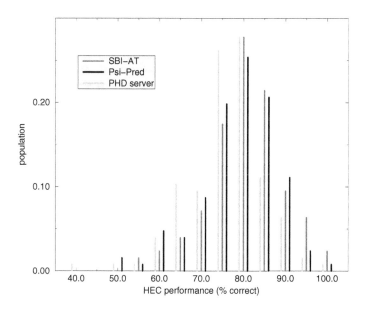

Fig. 2. Distribution of per-chain secondary structure prediction accuracy for SBI-AT (red), PSI-PRED (black) and PHD (green) methods.

Per-chain accuracies range from 55-100% (Figure 2) with an overall mean of 80.5%. Competitive prediction servers yield a mean of 78.8% (PSI-PRED, [7]) and 76.3% (PHD, [12, 13]).

4.1 Probability Transformation

HEC probabilities are obtained from network output activities via matrix transformation, which provides an additional 1.2% in performance over untransformed activities. A conversion matrix is obtained for each of 800 network combinations by processing the training data set.

4.2 Output Expansion

Output expansion is a procedure developed at SBI-AT which provides training hints to neural networks by predicting the secondary structure for more than one residue at a time. Predicting for three contiguous residues provides an additional 0.5% in performance over single-residue predictions.

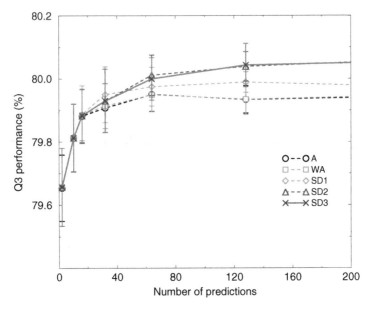

Fig. 3. Performance vs. Number of predictions and balloting threshold. Mean and standard deviations were obtained using random selections with 'replacement', which tends to underestimate performance.

4.3 Combinations of Many Predictions

Prediction accuracy increases asymptotically (Figure 3) as more predictions are combined to obtain HEC probabilities. Using more than 200 predictions provides an additional 0.2% in performance over using only 10 predictions [4].

An efficient balloting procedure was developed in which 'confident' predictions dominate in a weighted average of secondary structure probabilities.

References

1. Altschul, S. F., Gish, W., Miller, W., Myers, E. W. & Lipman, D. J. (1990) J. Mol. Biol. 215, 403–410.
2. Altschul, S. F., Madden, T. L., Schffer, A. A., Zhang, J., Zhang, Z., Miller, W. & Lipman, D. J. (1997) Nucleic Acids Res. 25, 3389–3402.
3. Bohr, H., Bohr, J. & Brunak, S., et al. (1988) FEBS Lett 241, 223–228.
4. Chandonia, J. M. & Karplus, M. (1999) Proteins 35, 293–306.
5. Chou, P. Y. & Fasman, G. D. (1974) Biochemistry 13, 211–222.
6. Cuff, J. A. & Barton, G. J. (1999) Proteins 34, 508–519.
7. Jones, D. T. (1999) J. Mol. Biol. 292, 195–202.
8. Kabsch, W. & Sander, C. (1983) Biopolymers 22, 2577–2637.

9. Lund, O., Frimand, K., Gorodkin, J., Bohr, H., Bohr, J., Hansen, J. & Brunak, S. (1997) Protein Engineering 10, 1241–1248.
10. Orengo, C.A., Bray, J.E., Hubbard T., LoConte L. & Sillitoe I. (1999) Proteins Suppl 3, 149–170.
11. Petersen, T. N., Lundegaard, C., Nielsen, M., Bohr, H., Bohr, J., Brunak, S., Gippert, G. P. & Lund, O. (2000) Proteins 41, 17–20.
12. Rost, B. & Sander, C. (1993) J. Mol. Biol. 232, 584–599.
13. Rost, B. & Sander, C. (1994) Proteins 19, 55–72.

Self-consistent Knowledge-Based Approach to Protein Design

Andrea Rossi[1], Cristian Micheletti[1], Flavio Seno[2], and Amos Maritan[1]

[1] SISSA, INFM and the Abdus Salam International Center, Trieste - Italy
[2] INFM - Dipartimento di Fisica, Università di Padova, Padova - Italy

1 Introduction

Two of the most investigated problems in molecular biology are protein folding and design. These problems stem from Anfinsen's discovery [1] that the sequence of amino acids of a naturally-occurring protein uniquely specifies its thermodynamically stable native structure. The protein folding challenge consists of predicting the native state of a protein from its sequence of amino acids, while in protein design one is concerned to identify the amino acid sequences folding into a pre-assigned native conformation. This last issue, having obvious practical and evolutionary significance, has attracted considerable attention and effort of experimentalists and theorists [2, 3, 4, 5, 6, 7, 8].

The difficulty of the protein design problem is enormous because, in principle, a rigorous approach [3, 7] would entail a simultaneous exploration of both the family of viable sequences and the family of physical conformations. By doing so, it would be possible to find the sequences having lower energy in the target structure than in any other conformation. Stated mathematically, to design a target structure Γ, one needs to identify the sequence of amino acids, s, that maximizes the "occupation probability" according to Boltzmann statistics:

$$P_s(\Gamma) = \frac{\exp\left(-\beta H_s(\Gamma)\right)}{\sum_{\{\Gamma'\}} \exp\left(-\beta H_s(\Gamma')\right)} = \frac{\exp\left(-\beta H_s(\Gamma)\right)}{Z_s} \tag{1}$$

evaluated at a suitable physiological temperature, $1/\beta = k_B T$. $\{\Gamma'\}$ denotes the family of conformations that can house the sequence s and $H_s(\Gamma')$ is the energy of the sequence in the conformation Γ'.

A first obstacle in using Eq. (1) is the difficulty of determining $H_s(\Gamma)$. However, even assuming the correct knowledge of H, it would be impossible to carry out an exhaustive search of the sequence maximizing $P_s(\Gamma)$, due to

C. Guerra, S. Istrail (Eds.): Protein Structure Analysis and Design, LNBI 2666, pp. 123-129, 2003.
© Springer-Verlag Berlin Heidelberg 2003

the computational difficulty of determining of Z_s accurately. By writing $Z_s = \exp(\log Z_s)$ and taking the first order term in its cumulant (high-temperature) expansion, the condition of maximizing P_s yields

$$W_s(\Gamma) \equiv H_s(\Gamma) - \langle H_s \rangle \ll 0 \; , \tag{2}$$

where the average $\langle \cdots \rangle$ is carried out over all the conformations Γ that can house s. Two main problems remain unsolved before the minimization of $W_s(\Gamma)$ in sequence space can be exploited in automated contexts to design protein structures. In fact neither the functional form of the effective energy H_s is known, nor its average, $\langle \cdots \rangle$, is easily computable being dependent on a large set of unknown conformations. In the next section we will introduce a plausible simple form for H_s and will determine the corresponding W_s by using data from a set of real proteins.

2 The Design Strategy

To represent a protein we shall use the common coarse-grained modelling where each amino acid is identified by a centroid placed on the β carbon (α carbon for Glycine).

Furthermore, we shall also partition the 20 types of amino acids into a restricted number of classes. More precisely we will adopt the partition where Ala, Ile, Leu, Met, Phe, Pro, Trp, and Val are in the same class (*hydrophbic*), Asn, Cys, Gln, Gly, Ser, Thr, and Tyr are in the class of the *neutral polar* and, finally, Arg, His, Lys, Asp, and Glu are in the *charged polar* class.

This simplification stems mainly from the observation that most amino acids in natural proteins can be substituted by "chemically equivalent" ones without disrupting native folds [9]. Hence, within the present design scheme we aim at predicting the classes of amino acids designing a given structure. As in ref. [8], the putative solution could then be fine-grained into 20 amino acids alphabet by using steric packing and solvation constraints.

Finally, we introduce a suitable (free) energy scoring function. The most popular choice adopted in simplified models is the pairwise-interaction form

$$H_s(\Gamma) = \sum_{i<j} \Delta_{ij}(\Gamma) B(s_i, s_j) \; , \tag{3}$$

where i, j are the positions along the sequence of the amino acids and the sum is taken over all possible pairs. $B(s_i, s_j)$ represents the interaction strength of the amino acid pair s_i and s_j. However, only amino acids that are close enough will interact in a non-negligible way. This is enforced with a suitable weight function, or contact map.

Due to the linear dependence of the energy H on the contact map (the only factors that contain geometric information about structures), the r.h.s. of Eq. (2) can be re-casted into the following forms:

$$\langle H_s \rangle = \sum_{i<j} \langle \Delta_{ij} \rangle B(s_i, s_j) \,, \tag{4}$$

The average contact map, $\langle \Delta(i,j) \rangle$, was obtained by collecting data on the contact map of a variety of naturally occurring proteins. It turns out that $\langle \Delta(i,j) \rangle$ is well reproduced as a (decreasing) function of the sequence separation, $|i - j|$ when $|i - j| < 16$. Contacts between residues with sequence separation larger than 16 are rather rare hence were modeled by assuming a constant frequency of occurrence, $\Delta^{(0)}$. The value of $\Delta^{(0)}$ is a free parameter that is to be tuned separately for each protein length so that the average number of overall contacts, $\sum_{i,j} \Delta_{i<j}$, matches the one observed in nature.

By using Eqs. (2) and (3), the scoring funcion W_s defined in Eq. (1) can be re-written as:

$$W_s(\Gamma) = \sum_{i<j} (\Delta_{ij}(\Gamma) - \langle \Delta_{ij} \rangle) B(s_i, s_j) \,, \tag{5}$$

where the $B(s_i, s_j)$ are parameters that have been set by requiring that the scores $W_s(\Gamma)$ associated with the native sequences of 30 proteins on their own native state be as small as possible, consistently with Eq. (2).

3 Results and Discussion

The design of a target structure, Γ, is carried out by minimizing the score, $W_s(\Gamma)$ over the possible sequences of amino acids classes. The exploration of sequence space is carried out within a stochastic scheme (simulated annealing, the elementary move being the random mutation of a fraction of residues from one class to another).

The minimization of $W_s(\Gamma)$ is achieved by varying a control parameter, T, that influences the rate at which sequences with lower and lower values of $W_s(\Gamma)$ are accepted. It is instructive to carry out our design scheme on naturally occurring target structures. By doing so, it would be possible to compare the putative design solution with the one adopted by nature. A possible complication comes from the fact that proteins with sequence identity as low as 30 % can share the same fold [10, 11]. After the coarse graining into the three classes mentioned before, this threshold value for homology becomes 55 %. Thus, at the simplest level, a match of about 55 % between our design solution and the one adopted by nature can be considered succesful. This value is remarkably close to the best design scores achieved with our procedure (data not shown). This does not imply automatically that our solutions are viable. Site-directed mutagenesis experiments [13] have shown that a few protein sites do not tolerate any substitutive mutation at all (otherwise the native state would be destabilized). It should then be checked whether such key residues, which are conserved in homologous proteins, are conserved also by our design strategy.

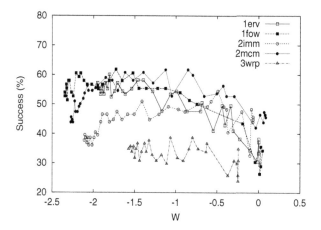

Fig. 1. The success as a function of the cost function $W(s, \Gamma_t) = H_s(\Gamma_t) - \langle H_s \rangle$ per site. Success is defined here by majority rule on a sampling of hundred (decorrelated) sequences. The value of the cost function for the respective wild-type sequences is between -0.48 and -0.78.

A further difficulty may stem from the fact that naturally occurring sequences may not have evolved to provide maximum thermodynamic stability, $W_s(\Gamma)$. Under these circumstances, one would observe that design solutions which extremize $W_s(\Gamma)$ give a worse match with the natural protein than sequences with a higher $W_s(\Gamma)$ score. For this reason we chose to test the success rate not only for the minimum value of $W(s)$, but also for other sequences. In particular it is interesting to compare all the sequences s with $W(s) < W(s^*)$, where s^* is the wild-type sequence. For each annealing temperature we extract 100 decorrelated sequences and make statistical analysis on this sequence set. We evaluate the average of $W(s)$ for this set and a "super-sequence" by applying a pointwise majority rule to this set: for each site we assign the most frequent amino acid class observed in this sequence set at the given location. It appears that, indeed, the highest matching with the native sequence, is not obtained for the lowest value of $W_s(\Gamma)$, but for higher ones as shown in Fig. 1 . Furthermore one can see (figure not shown, see [12] for details) that the sites that are assigned unambiguosly already at high values of $W_s(\Gamma)$ (i.e. early in our stochastic minimization), have a high probability to match the natural solution (for protein 1erv for the top 40 sites, there are 32 correct matches!). It is tempting to conjecture that the residues that are assigned with very little uncertainty by our design procedure (conserved design residues) could also correspond to conserved residues in nature, i.e. amino acids that play a fundamental role in the folding process.

This hypothesis is confirmed by the results of design attempts on two heavily investigated proteins: barnase and chymotrypsin inhibitor [13, 14, 15].

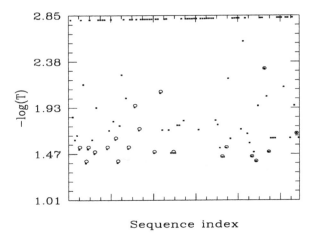

Fig. 2. Scatter plot of the threshold value of the control parameter, T, at which the designed class of each residue in barnase (sequentially labelled on the x axis) is assigned with uncertainty less than 50 %. Lower values on the y axis correspond to early locked sites in our design procedure. Circled dots represent sites belonging to *core1*, *core2* or *core3* .

For each of them several experimental results are available which pinpoint the key folding sites. For both proteins we found a striking agreement between the conserved residues in our approach and the ones identified in mutagenesis experiments. For example (see Fig. 2), for barnase, the residues for which the class-locking occurs at high values of the control parameter, T, (i.e. large values of $W_s(\Gamma)$) correlate very well with the hydrophobic *core1* which Fersht[14] identified as the initiator of the folding transition. Analogous results hold for CI2, where the top six conserved residues in our scheme contain the three residues (ALA35, ILE76, LEU68) indicated by Itzhaki *et al.*[15] as the most important in the folding process.

These striking results serve a two-fold purpose. On one hand they confirm the validity of the present design approach; on the other they also show some of its possible applications, in connection with the prediction of folding nucleus.

4 Summary

To summarize, we carried out automated protein design attempts over some PDB conformations by introducing several novel strategies to identify optimal energy-cost functions and select putative design solutions. A mere comparison of designed sequences with the PDB ones gives a success rate between 40% and 55% when working with three classes of amino acids: a value well above the random-guessing threshold. This success rate is not improving by introducing more sophisticated energy functions, suggesting that important features of real proteins are neglected by short range Hamiltonians. Nevertheless, a statistical analysis of a wider set of possible solutions, shows how the design procedure could be used to predict, with a high confidence, at least a sub-set of protein sites. These residues can be related to the conserved sites obtained by a statistical analysis of naturally occurring homologous sequences. Moreover, for two specific proteins (barnase and chymotrypsin inhibitor), these highly predictable sites correspond with a very good precision, to the folding nucleus, which is crucial for the folding process.

Acknowledgments This work was supported by INFM (PAIS project) and MURST (COFIN99).

References

1. C. Anfinsen. Principles that govern the folding of protein chains. *Science*, 181:223–239, 1973.
2. S. Sun, R. Brem, R. Chan, and K. Dill. Designing amino acid sequences to fold with good hydrophobic cores. *Protein Eng.*, 8:1205–1213, 1995.
3. F. Seno, M. Vendruscolo, J. Banavar, and A. Maritan. Optimal protein design procedure. *Phys. Rev. Lett.*, 77:1901–1904, 1996.
4. J. Deutsch and T. Kurosky. New algorithm for protein design. *Phys. Rev. Lett.*, 76:323–326, 1996.
5. B. Dahiyat and S. Mayo. De novo protein design: fully automated sequence selection. *Science*, 278(5335):82–87, 1997.
6. C. Micheletti, F. Seno, A. Maritan, and J. Banavar. Design of proteins with hydrophobic and polar amino acids. *Proteins: Structure Function and Genetics*, 32:80, 1998.
7. C. Micheletti, J. Banavar, A. Maritan, and F. Seno. Protein structures and optimal folding from a geometrical variational principle. *Phys. Rev. Lett.*, 82:3372–3375, 1999.
8. A. G. Street and L. S. Mayo. Computational protein design. *Structure with folding and design*, 7:R105–109, 1999.
9. S. Kamtekar, J. Schiffer, H. Xiong, J. Babik, and M. Hecht. Protein design by binary patterning of polar and nonpolar amino acids. *Science*, 262:1680–1685, 1993.
10. C. Chothia and A. M. Lesk. The relation between the divergence of sequences and structures in proteins. *EMBO J.*, 5:823–826, 1986.
11. C. Sander and R. Schneider. Database of homology-derived protein structures and the structural meaning of sequence alignment. *Proteins*, 9:56–68, 1991.
12. A. Rossi, C. Micheletti, F. Seno and A. Maritan Self-consistent knowledge-based approach to protein design. *Bioph. J.*, 80, 2001 (in press).
13. A. Fersht. *Structure and mechanism in proteinscience: a guide to enzyme catalysis and protein folding.* W.H. Freeman, New York, 1999.
14. A. R. Fersht. Optimization of rates of protein folding - the nucleation condensation mechanism and its implications. *Proc. Natl. Acad. Sci. USA*, 92:10869–10873, 1995.

15. L. S. Itzhaki, D. E. Otzen, and A. R. Fersht. The structure of the transition state for folding of chymotrypsin inhibitor 2 analysed by protein engeeniring methods: evidence for a nucleation-condensation mechanism for protein folding. *J. Mol. Biol.*, 254:260–288, 1995.

Protein Structure from Solid-State NMR

John R. Quine and Timothy A. Cross

Florida State University and National High Magnetic Field Laboratory

Summary. This article deals with mathematical questions arising from the determination of protein structure from data obtained by solid-state nuclear magnetic
resonance (NMR). Solid-state NMR holds the promise of revealing the structure of
membrane proteins in a lipid bilayer. The derivation of protein structure from NMR
data has most often been done using proteins in liquid state, and the mathematical
analysis has been done using distance geometry and distance matrices. The mathematical analysis for solid state NMR uses orientational constraints rather than
distance constraints, and matrices of inner products rather than distance matrices. Solving the structure from the data requires supplying a sequence of signs, a
situation somewhat analogous to the necessity to supply the phases to solve a structure from x-ray crystallographic data. Other problems in solving for the structure
arise from the condition that the gram determinants be non-negative, and this is
analogous problem in distance geometry that the distance matrix must satisfy the
conditions of the Cayley-Menger theorem.

1 Discrete Curves

It is convenient to think of a protein as a collection of discrete curves. A
discrete curve is a sequence of points $\mathbf{p}_0, \ldots, \mathbf{p}_n$ in three dimensional space,
which can be thought of as atomic coordinates. The backbone is naturally
a discrete curve consisting of points representing the atoms $-C_1-N-C_\alpha-C_1-$
proceeding from N-terminus to C-terminus. Although the atoms in side chains
have no natural sequential order, side chains are also formed from collection
of discrete curves.

A version of the Frenet frame for differentiable space curves can be given
for discrete space curves. Let

$$s_j = |\mathbf{p}_{j+1} - \mathbf{p}_j|$$

and define a unit tangent vector at $\mathbf{p}_j$, $j = 0, \ldots, n-1$, by

$$\mathbf{t}_j = \frac{\mathbf{p}_{j+1} - \mathbf{p}_j}{s_j}. \tag{1}$$

C. Guerra, S. Istrail (Eds.): Protein Structure Analysis and Design, LNBI 2666, pp. 131-137, 2003.
© Springer-Verlag Berlin Heidelberg 2003

The discrete curve can be reconstructed from the sequences $\{\mathbf{t}_j\}$ and $\{s_j\}$ by

$$\mathbf{p}_k - \mathbf{p}_1 = \sum_{j=1}^{k-1} s_j \mathbf{t}_j. \tag{2}$$

If $\mathbf{t}_{j-1}$ and $\mathbf{t}_j$ are not parallel, binormal and normal vectors can be given by

$$\mathbf{b}_j = \frac{\mathbf{t}_{j-1} \times \mathbf{t}_j}{|\mathbf{t}_{j-1} \times \mathbf{t}_j|} \qquad \mathbf{n}_j = \mathbf{b}_j \times \mathbf{t}_j \tag{3}$$

and a Frenet frame by

$$\mathbf{F}_j = (\mathbf{t}_j, \mathbf{n}_j, \mathbf{b}_j). \tag{4}$$

Generally, vectors will be thought of as column vectors, and frames as a sequence of three linearly independent column vectors considered as columns of a non-singular 3×3 matrix. Orthogonal frames correspond to orthogonal matrices and right-handed orthogonal frames to rotation matrices.

¿From crystallographic studies of small molecules [2], geometric parameters related to the discrete Frenet frame, such as the sequences $\{\mathbf{t}_{j-1} \cdot \mathbf{t}_j\}$ and $\{s_j\}$, are known to be independent of the particular protein, and dependent only on the types of atoms involved in the bond. Thus, bond angles C_1– N–C_α and C_1–C–N are approximately trigonal, 120°, and bond angles N–C_α–C_1 approximately tetrahedral, 109°, and bond lengths have standard values in the range 1 to 1.5 Angstroms, depending on the types of atoms in the bond.

These standard values of bond angles and lengths were of key importance in the initial investigations of protein secondary structure by Pauling and his co-workers. It is also of key importance in structure determination by solid state nuclear magnetic resonance (ss-NMR). The information available from ss-NMR experiments concerns the values of $\mathbf{t}_j \cdot \mathbf{B}$, i.e., the cosines of the angles of unit bond directions with the unit vector $\mathbf{B}$ giving the direction of the magnetic field. We refer to these as *bond direction cosines*. An equation in one or more bond direction cosines is referred to as an *orientational constraint*. Each observation gives an orientational constraint which is a quadratic equation in one or two of these bond direction cosines. Orientational constraints can be combined with information about the standard values for bond angles and bond lengths to get structural information about the protein.

Since the observation of hydrogens is a key NMR tool, bond angles involving hydrogen atoms are also important in ss-NMR structure determination, and standard values are used also for these bond angles involving hydrogens [6]. This is a major difference from x-ray crystallography where hydrogen atoms are not seen.

The discrete Frenet frames are also related to the usual torsion angles used in molecular structure. The relationship of one Frenet frame to the next is given by

$$\mathbf{F}_{j+1} = \mathbf{F}_j \, \mathbf{R}_3(\theta_{j+1}) \, \mathbf{R}_1(\tau_j) \tag{5}$$

where $\theta_j = -\arccos(\mathbf{t}_{j-1} \cdot \mathbf{t}_j)$, and τ_j is the torsion angle about the bond direction $\mathbf{t}_j$, and where

$$\mathbf{R}_1(\theta) = \begin{pmatrix} 1 & 0 & 0 \\ 0 & \cos\theta & -\sin\theta \\ 0 & \sin\theta & \cos\theta \end{pmatrix} \qquad \mathbf{R}_3(\theta) = \begin{pmatrix} \cos\theta & -\sin\theta & 0 \\ \sin\theta & \cos\theta & 0 \\ 0 & 0 & 1 \end{pmatrix} \qquad (6)$$

are rotations about the x and z axes respectively. Note that θ_j is the exterior bond angle at $\mathbf{p}_j$. Thus the discrete curve can be reconstructed up to a Euclidean motion from the sequences $\{s_j\}$, $\{\mathbf{t}_{j-1} \cdot \mathbf{t}_j\}$ and the torsion angles sequence $\{\tau_j\}$, the bond lengths, bond angles and torsion angles.

The discrete Frenet frame at $\mathbf{p}_j$ is related to one or more molecular frames that can be defined there. A molecular frame can be defined for an atom bonded to two others. Suppose atom B is bonded to atoms A and C, then a right-handed orthogonal molecular frame $\mathbf{M}$ can be constructed by setting

$$\mathbf{u}_1 = \frac{\mathbf{p}_A - \mathbf{p}_B}{|\mathbf{p}_A - \mathbf{p}_B|} \qquad \mathbf{u}_2 = \frac{\mathbf{p}_C - \mathbf{p}_B}{|\mathbf{p}_C - \mathbf{p}_B|} \qquad \mathbf{b} = \frac{\mathbf{u}_1 \times \mathbf{u}_2}{|\mathbf{u}_1 \times \mathbf{u}_2|} \qquad (7)$$

and forming the frame

$$\mathbf{M} = (\mathbf{u}_1, \mathbf{b} \times \mathbf{u}_1, \mathbf{b}). \qquad (8)$$

This molecular frame is especially useful if the electron cloud is considered to be symmetric the the plane of the three atoms, as is the case in the peptide plane. In this case the local chemical and electric properties, as they relate to NMR, are often assumed to be symmetric in the plane of $\mathbf{u}_1$ and $\mathbf{u}_2$.

The frame $\mathbf{M}$ is related to the Frenet frame for the sequence $\mathbf{p}_A$, $\mathbf{p}_B$ and $\mathbf{p}_C$ or the sequence $\mathbf{p}_C$, $\mathbf{p}_B$ by a rotation leaving the plane of $\mathbf{u}_1$ and $\mathbf{u}_2$ fixed.

2 Tensors and NMR

The physics of NMR is quantum mechanics. An NMR experiment is the observation of the precession of nuclear spins in the presence of a magnetic field. The equation for this precession is derived from the Zeeman Hamiltonian. A typical solid-state NMR observable σ is described by a symmetric tensor $\mathbf{T}$ so that $\sigma = \mathbf{B}'\mathbf{TB}$, where $\mathbf{B}$ is the unit direction of the magnetic field. The analytical expression for this tensor comes from a perturbation of the Zeeman Hamiltonian for the magnetic field, and only the second order perturbation is significant. This is why the observable is given by a quadratic function of the coordinates of $\mathbf{B}$.

In NMR the signal from an ensemble of molecules is observed. In solid-state NMR, the molecular frames are fixed with respect to the the direction $\mathbf{B}$, as opposed to liquid state NMR where the molecular frames are randomly oriented with respect to $\mathbf{B}$. In our solid-state NMR experiments, the molecules are held fixed in membrane bilayers that are pressed between glass plates [1]. Thus in ss-NMR, for each piece of data there is an ensemble of tensors fixed with respect to the magnetic field direction. This is in contrast to liquid state NMR, where for each piece of data there is an ensemble of tensors $\mathbf{T} = \mathbf{R'DR}$ where $\mathbf{D}$ is a fixed diagonal tensor and $\mathbf{R}$ is a random rotation with respect to the laboratory frame. This rotation is random function both in the time variable and in the statistical measure of the ensemble.

Because the NMR observable is a change in spin precession frequency principally due to the chemical environment of the observed nucleus, each of these tensors can be thought of as fixed in a molecular frame. Thus $\mathbf{M'TM}$ is constant and can be determined by *powder experiments* where the ensemble of tensors is kept fixed with respect to time, but random with respect to the statistical measure of the ensemble.

Generally, one of the principal axes of $\mathbf{T}$ is aligned with $\mathbf{u}_1 \times \mathbf{u}_2$ and

$$\mathbf{M'TM} = \mathbf{R}_3(\beta)\, \mathrm{diag}(\sigma_1, \sigma_2, \sigma_3)\, \mathbf{R}_3(\beta)' \tag{9}$$

where diag indicates a diagonal matrix and σ_1, σ_2, and σ_3 are the principal values of the tensor. The principal values and the parameter β are determined from the powder experiment. A consequence of (9) is that for fixed bond angle, and hence fixed $\mathbf{u}_1 \cdot \mathbf{u}_2$, the NMR observable $\mathbf{B'TB}$ is a quadratic in $\mathbf{u}_1 \cdot \mathbf{B}$ and $\mathbf{u}_2 \cdot \mathbf{B}$, and setting it equal to its observed value gives and orientatioal constraint.

The most common observables in our ss - NMR experiments are the quadrupolar splitting, the dipolar splitting, and the chemical shift. The quadrupolar and dipolar splittings give rise to zonal harmonic tensors with $2\sigma_1 = -\sigma_2 = -\sigma_3$, and for these there is a unique principal major axis along a bond direction. For the chemical shift tensors, the principal values are usually distinct with σ_1 the largest value, so that the maximum chemical shift occurs in the $\mathbf{u}_1$, $\mathbf{u}_2$ plane of the molecular frame.

3 Structure from Orientational Constraints

We turn to the question of whether it is possible to solve for a structure from orientational constraints. In fact, the analytical solution includes some undetermined signs, ± 1. The situation is somewhat analogous to crystallography, where the analytic solution for the electron density map as an inverse Fourier transform includes a collection of undetermined absolute value 1 complex numbers, or phases.

As shown above, orientational constraints come from the expression of NMR observables as quadratic tensors, giving quadratic equations in the bond-orientation cosines $\{\mathbf{B} \cdot \mathbf{t}_j\}$ where $\{\mathbf{t}_j\}$ are unit tangent vectors along a discrete curve and $\mathbf{B}$ is the unit direction of the magnetic field. With enough of this kind of data, the quadratic equations can be solved up to undetermined ± 1 signs coming from quadratic formula. We refer to these signs as *quadratic indices*

Let

$$\kappa_j = \mathbf{B} \cdot \mathbf{t}_j \qquad \beta_j = \mathbf{t}_{j-1} \cdot \mathbf{t}_j \tag{10}$$

Proceeding under the assumption that the quadratic indices are known and the sequence κ_j has been found, we solve for the unit vectors $\{\mathbf{t}_j\}$. Once the unit tangent vectors are found, the atom coordinates can be found from the sequence $\{s_j\}$ of bond lengths using (2).

We can solve recursively for $\{\mathbf{t}_j\}$ under the condition that $\kappa_j^2 \neq 1$ for all j. Recall that we also assume that the bond angles are known, so that the sequence β_j is known.

To solve for the sequence $\{\mathbf{t}_j\}$, let

$$G_j = \begin{pmatrix} 1 & \kappa_{j-1} & \kappa_j \\ \kappa_{j-1} & 1 & \beta_j \\ \kappa_j & \beta_j & 1 \end{pmatrix} \tag{11}$$

be the matrix of inner products (gram matrix) of the vectors $(\mathbf{B}, \mathbf{t}_{j-1}, \mathbf{t}_j)$. Let

$$g_j = \det G_j = 1 - \kappa_{j-1}^2 - \kappa_j^2 - \beta_j^2 + 2\kappa_{j-1}\kappa_j\beta_j, \tag{12}$$

be the sequence of determinants. Now choose a lab frame $(\mathbf{i}, \mathbf{j}, \mathbf{k})$ with $\mathbf{B} = \mathbf{k}$, then check that

$$\mathbf{t}_0 = \sqrt{1 - \kappa_0^2}\,\mathbf{i} + \kappa_0\,\mathbf{k}$$

$$\mathbf{t}_j = \frac{1}{(1 - \kappa_{j-1}^2)} \left((\beta_j - \kappa_j\kappa_{j-1})\,\mathbf{t}_{j-1} + (\kappa_j - \beta_j\kappa_{j-1})\,\mathbf{k} + \varepsilon_j\sqrt{g_j}\,\mathbf{t}_{j-1} \times \mathbf{k} \right).$$

$$\tag{13}$$

where $\varepsilon_j = \pm 1$ solves (10) recursively for the sequence $\{\mathbf{t}_j\}$ of unit vectors [7]. Note that

$$\mathbf{B} \cdot (\mathbf{t}_j \times \mathbf{t}_{j-1}) = \varepsilon_j\sqrt{g_j}, \tag{14}$$

i.e., ε_j is the sign of the above triple product. The ε_j are referred to as *chiralities*.

The requirements for solving for the sequence $\{\mathbf{t}_j\}$ using (13) are

1. $\kappa_j^2 \neq 1$ for $j = 1, \ldots, n$
2. $g_j \geq 0$ for $j = 1, \ldots, n - 1$,
3. the choice of a sequence ε_j of chiralities.

Each of these requirements can lead to some problems in practice. The first requirement can be a problem if one of the unit bond vectors to be determined is parallel to $\mathbf{B}$. In this case the sequence $\{\mathbf{t}_j\}$ cannot be determined uniquely. If, for example, one unit tangent vector is parallel to $\mathbf{B}$ then all of the subsequent unit vectors can be rotated about $\mathbf{B}$ by the same rotation without changing the bond angles or orientation constraints.

The second requirement is due to the fact that the gramian determinant of three linearly independent vectors must be positive definite, since it gives the metric tensor. If some of the g_j are computed to be negative, then some of the data is bad, or the assumption about the bond angle is bad. The situation is analogous to distance geometry where the distance matrix must be positive semi-definite by the Cayley-Menger Theorem [3].

The third requirement indicates again there is another choice of signs required to solve for the structure from the orientational constraints.

The method of dealing with these problems varies. Occasionally the data is good enough so that a sequence of orientations and bond angles can be supplied so that 1. and 2. above are satisfied. The chiralities and the quadratic indices must then be supplied and an initial structure obtained. There may be a set of reasonable choices for these indices and several intitial structures. These structures should then be refined using an energy penalty and a stereochemical energy function in a way analogous to x-ray crystallographic refinement [4] [5].

4 Acknowledgment

This work was supported by the National Science Foundation, DMB 9986036 (JRQ and TAC), DMS 9972858 (JRQ). This work was largely performed at the National High Magnetic Field Laboratory supported by NSF Cooperative Agreement (DMR 9527035) and the State of Florida.

References

1. T. A. Cross and J. R. Quine. Protein structure in anisotropic environments: development of orientational constraints. *Concepts in Magnetic Resonance*, 12, 2000.

2. R. A. Engh and R. Huber. Accurate bond and angle parameters for x–ray protein structure refinement. *Acta Crystallogr. A*, 47:392–400, 1991.

3. T. F. Havel and A. W. M. Dress. Distance geometry and geometric algebra. *Foundations of Physics*, 23:1357–1374, 1993.

4. R. R. Ketchem, K.-C. Lee, S. Huo, and T. A. Cross. Macromolecular structural elucidation with solid-state NMR–derived orientational constraints. *J. Biomol. NMR*, 8:1–14, 1996.

5. R. R. Ketchum, B. Roux, and T. A. Cross. High-resolution polypeptide structure in a lamellar phase lipid environment from solid state NMR derived orientational constraints. *Structure*, 5:1655–1669, 1997.

6. A. Kvick, A.R. Al-Karaaghouli, and T.F. Koetzle. Deformation electron density of α- glycylglycine at 82 K. I. the neutron diffraction study. *Acta Crystallogr.*, B33:3796–3801, 1977.

7. J. R. Quine and T. A. Cross. Protein structure in anisotropic environments: unique structural fold from orientational constraints. *Concepts in Magnetic Resonance*, 12, 2000.

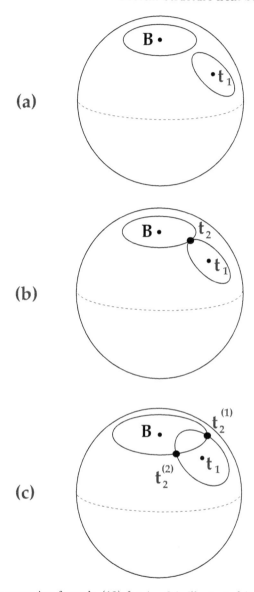

Fig. 1. The recursion formula (13) for $j = 2$ is illustrated in the figure. The sphere represents the set of unit vectors. The equations $\kappa_2 = \mathbf{B} \cdot \mathbf{t}_2$ and $\beta_2 = \mathbf{t}_1 \cdot \mathbf{t}_2$ indicate that given $\mathbf{t}_1$, the the vector $\mathbf{t}_2$ lies on both of two circles about $\mathbf{B}$ and $\mathbf{t}_1$.

a) If $g_2 < 0$ the circles do not intersect and there is no solution.
b) If $g_2 = 0$ then the circles intersect at one point and there is only one possibility for $\mathbf{t}_2$.
c) If $g_2 > 0$ then there are two possibilities for $\mathbf{t}_2$ distinguished by the chirality ε_2. The vector $\mathbf{t}_2^{(1)}$ corresponds to $\varepsilon_2 = -1$ and the vector $\mathbf{t}_2^{(2)}$ corresponds to $\varepsilon_2 = 1$.

Learning Effective Amino-Acid Interactions

Flavio Seno[1], Cristian Micheletti[2], Amos Maritan[2], and
Jayanth R. Banavar[3]

[1] INFM - Dipartimento di Fisica, Università di Padova, Padova - Italy
[2] SISSA, INFM and Abdus Salam Centre for Theoretical Physics, Trieste - Italy
[3] Department of Physics, Penn-State University, University Park -USA

1 Introduction

1.1 Background

Proteins are heteropolymers of 20 amino acids with genetically determined sequences. Most proteins assume specific globular conformations under physiological conditions. Several experiments indicate that this unique native state is thermodynamically stable and encoded unambiguously by the sequence of amino acids [1].

Two challenging problems arise from these facts: protein folding and protein design. Protein folding consists of predicting the native state from the known sequence of amino acids; protein design addresses the possibility of predicting which sequence admits a pre-assigned biologically viable conformation as its native state[1]. Tackling these problems with a "first principles" entailing the correct treatment of the quantum mechanical behaviour of the large number of atoms constituting a protein and its surrounding solvent is beyond the present computational capabilities. A natural way to avoid dealing with too many microscopic degrees of freedom is to introduce a reduced representations where each amino acid is represented by one or a few interaction sites. The main difficulty with such simplified representations is the introduction of an effective energy function that captures the essential qualitative physical and chemical features of the folding process. This involves both choosing the right form for the energy function as well as determining the precise value of the coarse-grained interaction potentials between the amino acids.

In this study we describe a method that allows to determine of these potentials and to verify the reliability of the chosen parametrisation of the energy function.

C. Guerra, S. Istrail (Eds.): Protein Structure Analysis and Design, LNBI 2666, pp. 139-145, 2003.

1.2 Determination of Interaction Potentials

Let us consider a coarse-grained description of a protein whose energy is described by an unspecified effective energy function depending on a set of p parameters (e.g. interaction potentials, chemical potentials etc.) $(\alpha_1, \ldots, \alpha_p)$. The particular parametrisation is crucial: one would like to capture the key ingredients such as size and steric constraints, the polar or hydrophobic nature of the amino acids, electric charge and hydrogen bonding with as few parameters as possible. Assuming that a physical parametrisation is available, one could extract the parameters by observing[2, 3] that, given a protein sequence S, the correct set of $\{\bar{\alpha}\}$ ought to recognize its native state Γ as having lower energy than any arbitrary, though physically viable and distinct from Γ, conformation Γ^{d} of the same length (decoy conformations), namely:

$$H(S, \Gamma, \bar{\alpha}) < H(S, \Gamma^{\mathrm{d}}, \bar{\alpha}) \quad \Gamma \neq \Gamma^{\mathrm{d}} , \tag{1}$$

where $H(S, \Gamma, \alpha)$ is the energy of the sequence S mounted on Γ obtained with the energy function H and the parameters $\{\alpha\}$. In other words, the correct potentials must be chosen from the region of parameter space satisfying inequalities (1) for any choice of viable decoys, $\{\Gamma^d\}$.

The Protein Data Bank (PDB) contains the native state structures of a large number of amino acid sequences. For each of them we could, in principle, collect a large set of inequalities such as (1) as long as we are able to find suitable decoy conformations. The goal is then to infer the energy function and the set of parameters that satisfy these inequalities for all native states and associated decoys. These scheme has the appealing feature of showing whether the chosen parametrisation of H can be appropriate or not. In fact, the impossibility of finding parameters recognizing the native states as the lowest energy ones with respect to the decoys would signal that the form for the energy function is unsuitable. The key difficulty to implement this approach is the generation of relevant decoys, i.e. structures that are really competitive with the native one to be ground-states of S. In the next section we introduce a new procedure to generate significant competitive structures.

1.3 An Iterative Strategy

The key idea [4] of our method is the observation that the native states of proteins must at least satisfy the requirement of being located at the bottom of a smooth free energy minimum with a wide basin of attraction [5, 6] . This suggests a straightforward approach:
a) begin with an initial guess of the potential parameters;
b) start from the native states of several proteins and carry out an unbiased Monte Carlo (or molecular dynamics) simulation (say at zero temperature) and determine several accessible local minima for each of the proteins.
c) Modify the potential parameters in such a way as to destabilize these conformations in favour of the known native state, according to eq. (1).
d) Iterate this procedure by returning to (b).

After several iterations, one would expect to converge to a set of potential parameters which best capture the shape of the free energy landscape in the vicinity of all the native state structures. Our method adopts the thermodynamic stability scheme described before but with the proviso that the decoys would be generated by an explicit and simple dynamical process. The structures so generated are guaranteed to be stringent competitors of the native structure. Since the scheme is both flexible and optimal, it can be used to compare the performance of many different scoring functions or parametrizations and hence select the most promising one for *ab initio* folding simulations. There are obviously many ways of implementing this idea and we will present a few schemes here that we have tried and which yield remarkably good results.

2 Models and Techniques

2.1 The Model and the Dynamics

We have used a highly simplified model of a protein in which the amino acid residue backbone is represented by a self-avoiding chain of connected C^α atoms. The C^α positions of a real protein are mapped on suitably chosen sites of an FCC lattice[7, 8] with unit spacing equal to the typical separation of subsequent C^α atoms (3.8 Å). This framework (XFCC model) allows for a faithful representation of protein backbones, since the coarse-grained C^α positions are typically within 1 Å of the crystallographic positions [7]. The choice of the proper FCC sites is done at the expenses of slight variations in the peptide bond lengths but it allows to preserve the typical torsion angles found in real protein molecules. For each amino acid we then define also the position of the side centroid (C^β atom) using standard geometrical rules[9].

We used a pairwise energy function, involving only interactions between pairs of amino acids:

$$\mathcal{H} = \sum \Delta(r_{ij}) \cdot \epsilon(S_i, S_j) + 10 \cdot \epsilon_r \cdot \left[\left(\frac{4.65}{r_{ij}} \right)^2 - 1 \right] \cdot \Omega(r_{ij}) \qquad (2)$$

$$\text{where}: \quad \Omega(r) = \Theta(4.65 - r) \qquad \Delta(r) = \frac{1}{2} + \frac{1}{2} \tanh \frac{6.5 - r}{2} \qquad (3)$$

Θ is the step function and $\mathbf{r}_{ij}$ is the distance (in Angstroms) between the C_α atoms of the i-th and j-th amino acid. Δ denotes the distance-dependent strength of interactions between the i-th and the j-th amino acids along the sequence mounted on the structure Γ, ϵ is the interaction matrix and S_i denotes the type of the ith amino acid in the sequence. Finally, ϵ_r is a repulsive term that penalizes cases where two non-consecutive pairs of C^α's are closer than 4.65 Å.

The dynamics in conformation space is carried out using a Monte Carlo technique. A new conformation is accepted according to the Metropolis rule. At each attempted MC step, we move up to 2 of consecutive protein residues to unoccupied positions. The new positions have to satisfy suitable physical constraints that we deduced by the analysis of the CA and CB positions of an ensemble of over hundred single-chain proteins. More precisely: the separation d between two consecutive C^{α} atoms (measured in Å) must remain in the range $2.6 < d < 4.7$; two non-consecutive C^{β} atoms must not be closer than 2 Å; two non-consecutive C^{α} and C^{β} atoms must not be closer than 2 Å; the chain length is allowed to fluctuate by up to a maximum value of 4 Å.

2.2 Perceptron Learning

A convenient way to find the optimal potentials (given the decoys confor-mation) is the use of the perceptron algorithm [10] for the optimization of a set of linear inequalities. In our case, the inequalities are of the form: $\mathcal{H}(S, \Gamma) - \mathcal{H}(S, \Gamma_d) < 0$. This expression can be rewritten as

$$\sum_{i>j=1}^{20} (n_{ij}^d - n_{ij}^n)\epsilon(i,j) + \epsilon_r(n_r^d - n_r^n) \equiv \sum_{i>j=1}^{20} a_{ij}(\Gamma_d) \cdot \epsilon(i,j) + a_r \cdot \epsilon_r$$
$$\equiv \mathcal{Q}(\Gamma_d, \epsilon) > 0 . \tag{4}$$

where $n_{ij}^{n,d}$ denotes the number of native/decoy contacts involving amino acids of types i and j and $n_r^{n,d}$ denotes the strength of the native/decoy repulsive term. Given the native state Γ and the sequence S, the 211 entries of $a_{i,j}(\Gamma_d)$ plus a_r depend on the geometrical properties of the decoys. For a given set of M inequalities to be satisfied simultaneously, it is convenient to identify the one (denoted with l) that, with the trial potentials is the worst satisfied one:

$$\mathcal{Q}(\Gamma_l, \epsilon) < \mathcal{Q}(\Gamma_k, \epsilon) \qquad k = 1, \ldots, M, \quad k \neq m \tag{5}$$

The selection of the conformation l can be done both when $\mathcal{Q}(\Gamma_l, \epsilon)$ is negative (not all the inequalities are satisfied) and when it is positive (all the inequalities are satisfied). Once l has been determined, one updates the trial potentials, $\epsilon(i,j)$ (or ϵ_r) by adding a quantity proportional to $(a_{ij}(\Gamma_l))$ (or $a_r(\Gamma_l)$. With these new potentials, each inequality is re-valuated and the cycle repeated. This method can be shown to converge to the optimal solu-tion: $\mathcal{Q}(\Gamma_l, \epsilon)$ reaches a constant value [10], which can be of either sign. If it is negative, no set of potentials can be found that consistently satisfies all inequalities in the set (unlearnable problem).

3 Results

We began by considering a single protein, PDB code: 1vcc, which has 77 residues. Starting from a set of random potentials, we generated 30 decoys for which we computed the average root mean square deviation (RMSD),$\bar{g}$ native structure and its variance $\Delta\bar{g}$. With these decoys, we found the potentials by applying the perceptron algorithm. With the new interaction parameters, we generated 30 more decoys and kept repeating this procedure.

In Fig. 1 we show the $\bar{g}$, as a function of the number of iterations. It is remarkable that, with such a simplified model, $\bar{g}$ can be decreased dramatically from the initial value of about 6 Å to around 1.5-2.0 Å, which is just over the order of the experimental uncertainty! This provides a nice demonstration of the fact that it is possible to stabilize the native state in the native basin within a low uncertainty.

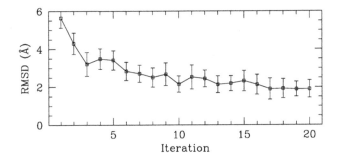

Fig. 1. Results of the iterations on the single protein 1vcc: we show the asymptotic RMSD of the decoy structures as a function of iteration.

Next, we attempted a task considerably more difficult: to stabilize the native states of twenty proteins simultaneously. The proteins were chosen among a list of non-redundant representatives of the main protein folds.

At each iteration step, we generated 5 decoys for each of them. We saw an improvement as the iterations went on, although not as pronounced as for the single 1vcc protein with $\bar{g}$ decreasing to a value of 3.8 ± 0.5 Å.

As recommended by Lazaridis and Karplus [11] as an independent test of the quality of our potentials we assessed their performance on a set of seven proteins (PDB codes: 1ctf, 1r69, 1sn3, 2cro, 3icb, 4pti, 4rxn) unrelated to those used for extracting the potentials and for which more than 600 stringent decoys structures (for each protein) have been derived[9]. This study ought to reflect the portability of our potentials, i.e. their applicability in contexts different from which they have been derived. For each protein we compute the energy of the the native state E_g and the energy of all the decoys E_i $(i = 1, \ldots, M)$ (where M is the number of decoys for each single protein) by using our potentials. With the correct potentials, E_g should be always lower than any other E_i. The native state is always among the best 5 to 10 %. This is a highly non-trivial result since it is generally difficult to get such correlations even employing specially designed energy scoring functions [9].

As a final test we introduced a slightly more sophisticated model where we consider interactions between all possible pairs of CA and CB. Using this second model, we get a clear improvement[4] of the results. This example is helpful in illustrating the possibility of using our novel optimization technique in selecting the "most physical" energy parameterization.

4 Conclusions

We have demonstrated how one may extract effective interaction potentials between amino acids in a coarse-grained description of a protein. The method relies on the possibility of finding a set of competitive decoys of the native state. We outlined an iterative procedure to generate these decoys which attempts to stabilize, at least locally, the native state. The results obtained with simple forms of the energy function are very promising – we were able to stabilize a set of 20 proteins to an average distance of less than 4 Å and moreover, the potentials, when tested on completely unrelated decoys, yield results which are remarkably good especially in view of the simplicity of the approach.

Acknowledgements.
This article was supported by grants from INFM and MURST-COFIN99.

References

1. T.E. Creighton: *Proteins: structure and molecular properties.* W.H. Freeman ed., New York 1992
2. G.M. Crippen: Prediction of protein folding from amino acid sequence over discrete conformation space. Biochemistry **30** (1991) 4232-4237
3. F. Seno, A. Maritan and J.R. Banavar: Interaction potentials for protein folding. Proteins: Structures, Function and Genetics **30** (1998) 244-248
4. C. Micheletti, F. Seno, A,. Maritan & J.R. Banavar: Learning effective amino acid interaction through iterative stochastic technique. In press Proteins: Function, Structure and Genetics (2000).
5. Bryngelson J. D. and Wolynes, P. G: Spin glasses and the statistical mechanics of protein folding, Proc. Natl. Acad. Sci. USA **84** (1987) 7524-7528
6. C. Micheletti, J. R. Banavar, A. Maritan and F. Seno: Protein structures and optimal folding from a geometrical variational principle. Phys. Rev. Lett. **82** (1999) 3372-3375
7. D.G. Covell and R. Jernigan: Conformations of folded proteins in restricted spaces. Biochemistry **19** (1990) 3287
8. B.H. Park and M. Levitt: The complexity and accuracy of discrete state models of protein structure. J. Mol. Biol. **249** (1995) 493-507
9. B.H. Park and M. Levitt: Energy functions that discriminate X-ray and near-native folds from well-constructed decoys. J. Mol. Biol. **258** (1996) 367-392

10. W. Krauth & M. Mezard: Learning algorithms with optimal stability in neural networks. J. Phys. **A20** (1987) L745-L752
11. T. Lazaridis and M. Karplus: Effective energy functions for protein structure prediction. Curr. Op. in Struct. Biol. **10** (2000) 139-145

Proteinlike Properties of Simple Models

Yves-Henri Sanejouand and Georges Trinquier

CRPP, Avenue Albert Schweitzer, 33600 Pessac, and IRSAMC, Université Paul Sabatier, 118 route de Narbonne, 31062 Toulouse Cédex, France.

During the last decade, it has been shown that several properties of natural proteins can be captured by simple models, such as 2-D or 3-D lattice models [1, 2, 5, 6, 9, 11, 13]. In these models, the protein is figured as a chain of beads occupying the sites of a lattice in a self avoiding way.

1 The 3x3x3 Cubic Lattice Model

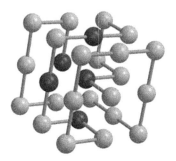

Fig. 1. One of the 103346 possible compact conformations of sequence PHP$_4$HPHPHP$_{15}$H. The hydrophobic (H) residues are the darkest ones.

In this section, we will focuss on some of the proteinlike properties exhibited by the compact conformations of chains of 27 monomers located at all sites of a $3 \times 3 \times 3$ cubic lattice, when only two kinds of monomers are considered,

C. Guerra, S. Istrail (Eds.): Protein Structure Analysis and Design, LNBI 2666, pp. 147-153, 2003.
© Springer-Verlag Berlin Heidelberg 2003

either hydrophobic (H) or polar (P) ones. Within the frame of this model, there are $2^{27} \simeq 1.3.10^8$ sequences and 103346 conformations[2], that is, 103346 different ways for a chain to go through all 27 lattice sites, going from a site to a neighboring one at each step. One of these so-called hamiltonian paths is shown in Fig. 1, for the following sequence: $PHP_4HPHPHP_{15}H$. Of course, Fig. 1 can also be viewed as showing another possible conformation, the reverse-labeled one, for the corresponding reverse-labeled sequence, namely, $HP_{15}HPHPHP_4HP$. If such pairs of conformations are assumed to be identical, the number of different compact conformations drops to 51704[6].

The fact that a sequence can be "threaded" in a given structure in two different ways, the position of the first residue in the first way being the position of the last residue in the other way, is certainly not a property of natural proteins, which are made with asymetric building blocks, namely, amino acids of the L-series. As a consequence, the $\alpha-$helix, one of their major structural elements, is right-handed, and left-handed helices have not yet been observed in the tridimensional structures of natural proteins. Note that the reverse-labeled sequence of a given protein could quite well be synthesized. If buildt with amino acids of the D-series, its tridimensional structure should be the same as that of the natural protein, as far as the positioning of the amino acid sidechains is concerned, except for proline residues, whose sidechain is involved in a five-member ring with the backbone. However, while complete syntheses of all-D proteins have been performed, to our knowledge, up to now, following the seminal work of Shemyakin and its collaborators [12], only small reverse-labeled all-D peptides, now called "retro-inverso" peptides[3], have been synthesized.

In order to study the sequence-structure relationship in the case of a given model, it is necessary to choose an energy function. For lattice models, the most usual one has the following form:

$$H = \sum_{i<j} E_{ij}\Delta(r_i - r_j)$$

where $\Delta(r_i - r_j) = 1$ if monomers i and j are close neigbors in the lattice, and $\Delta(r_i - r_j) = 0$ otherwise, and where E_{ij} depends on the nature of the interacting monomers. A popular choice for the E_{ij} values has been $E_{ij} = E_{HH} = -\epsilon$, when monomers i and j are both hydrophobic, and $E_{ij} = 0$ otherwise[2]. However, it has been noted that the effective interaction between amino acids in natural proteins is nearly additive[7], that is:
$$E_{ij} \simeq E_i + E_j$$

where E_i is a characteristics of the amino acids, strongly correlated with their hydrophobicity. When the strictly additive case is considered, with for instance $E_i = E_H = -1$ and $E_i = E_P = 0$, the sequence-structure relationship in the model exhibits some remarkable features. Noteworthy, if the energy is determined for all sequences and for all 103346 conformations, it is found that only 122750 of the sequences (0.09%) have non-degenerate ground states, that

Table I. 60 remarkable sequences of the $3 \times 3 \times 3$ cubic lattice model. When the additive potential is used, the ground state of each of them is one of the remarkable structures known for this model. The 60 remaining remarkable sequences are the reverse-labelled of these ones. These 120 sequences are the only ones with five hydrophobic residues and a non-degenerate ground state. For the sake of clarity, P residues are indicated by dots.

...............H.H.H.H..H.	...H..H.H.........H.H....
.............H.H...H.H..H.	...H..H.H.........H.....H..
...........H.....H.H.H..H.	...H..H.H.........H.H......
.........H.......H.H.H..H.	...H..H.H.......H.H........
.........H.....H...H.H..H.	...H..H.H....H...H........
........H...H.H.H..H.....	...H..H.H...H.............H
........H.H.......H.H..H.	...H..H.H.H.........H......
........H.H..H......H...H	...H..H.H.H.....H..........
........H.H..H....H.....H	..H.............H.H.H...H.
........H.H.H.....H....H.	..H...........H...H.H.H...
........H.H.H..H......H.	..H...........H.H..H...H..
........H.H.H..H..H.....	..H......H..H...H.H........
........H.H.H.H........H.	..H.....H...........H.H..H.
.......H.....H.H.H..H.....	..H.....H......H..H.H......
.......H.H..H........H...H	..H.....H....H...H.H........
.......H.H..H......H.H....	..H.....H..H..H.H..........
.......H.H..H....H.......H	..H.....H.H.H..........H.
.......H.H.H..H......H....	..H....H.............H.H.H
.......H.H.H.H......H.....	..H....H...H...H.H..........
......H..H.H.H.H..........	..H....H...H.H.............H
.....H.......H.H.H..H.....	..H...H.............H.H..H.
..H.............H...H.H.H	..H...H....H...H.H..........
..H.....H.....H.H.H......	..H..H...H.H.H..............
..H....H.............H.H.H	..H.H...............H.H..H.
..H....H.....H.H.H.......	..H.H..............H.H..H...
..H....H.H.H.......H......	..H.H.H.H................H.
..H..H.............H.H.H	.H................H.H.H...H
..H..H.......H.H.H.......	.H.............H.H.H......H
..H..H.H.............H.H	.H........H.H.H...........H
..H..H.H..............H...H	.H....H.H.H...............H

is, for each of them, a given conformation is the lowest-energy one, while, for all of them, Δ, the energy gap between this conformation and the second lowest-energy one, is: $\Delta = 2$. Strikingly, out of the 103346 possible conformations, only 120 (0.11%) are found to be possible ground states. All of these so-called remarkable structures are characterized by a large "designability"[6], as measured by N_s, the number of sequences they are the ground state of, N_s ranging between 513 and 2306.

In Fig. 1 the top structure[6] is shown, that is, the conformation which is the ground state of the largest number of sequences. Among these 2306 sequences, $PHP_4HPHPHP_{15}H$ has the smallest count of hydrophobic residues. As a matter of fact, each remarkable structure is the ground state conformation of a *single* five-hydrophobic residues sequence (all given in Table I), and the corresponding five residues are always located as shown in Fig. 1, that is, one of them being at the cube center, the four other ones being at the center of facets that are non bonded to the residue at the cube center. Furthermore, in all sequences sharing a given ground state, these five residues are always hydrophobic[13]. The latter point helps to clarify why there are 120 remarkable conformations. Indeed, if a sequence like $PHP_4HPHPHP_{15}H$ has a non-degenerate ground state, this means that, out of the 103346 possible ones, there is only one way to bring its five hydrophobic residues close together, so that each of them can interact with at least another hydrophobic one[13]. In other words, it is from a topological point of view that the 120 remarkable conformations of the $3 \times 3 \times 3$ cubic lattice model are atypical[8]. This is a quite satisfactory property of the model, since it means that the same 120 conformations are also expected to be remarkable when different E_{ij} values are chosen. For instance, with $E_{ij} = E_{HH} = -2 - \gamma, E_{ij} = E_{HP} = -1, E_{ij} = E_{PP} = 0$, and $\gamma = 0.3$, the conformation shown in Fig. 1 is still the top structure, but with $N_s = 3794[6]$, and the picture is now more complicated than in the case of the additive potential. The difference between the additive and nearly additive cases comes from the fact that the departure from additivity lifts the degeneracy of many sequences. Now energy gaps of $\Delta = 0 + n\gamma$ and $\Delta = 2 \pm n\gamma$ are observed, with $n = 1, 2$, *etc*[4]. For instance, with $\gamma = 0.3$, nearly half of the sequences whose ground state is one of the remarkable conformations have energy gaps lower than 1.0[13]. Moreover, while 4.75% of the sequences have a unique ground state, for the vast majority of them it is not one of the remarkable conformations, and the corresponding energy gap is, on average, close to 0.3, *i.e.*, the γ value[6]. When only sequences with energy gaps larger than 1.0 are considered, the picture obtained with the additive potential is restored, despite a few minor differences arising from the overlap of the energy gaps splitting around the $\Delta = 0$ and $\Delta = 2$ cases[13].

An interesting proteinlike property of the model allows for the determination of *all* large gap sequences (*i.e.*, with $\Delta = 2$, in the case of the additive potential) whose ground state is a given remarkable conformation, without the need of any huge enumeration[13]. To do so, starting for instance from the corresponding remarkable five hydrophobic residues sequence given in Table I, all 27 singly-mutated sequences are generated. Then, among them, those with the same single ground state are retained. Next, for each sequence of this subset, all 27 singly-mutated sequences are generated, and so on, until no new sequence can be retained. What is shown through the success of such a protocol is that the N_s sequences of each of the 120 remarkable structures belong to a "neutral island" of the sequence space[13]. Note that such a property has also been found in the case of square lattice models[1].

2 N-Soft-Spheres Models

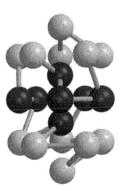

Fig. 2. One of the possible compact conformations of sequence $P_2H_2P_4H_3P_4H_2P_2$. Note that such a geometry is close to the cubo-octahedral one found to be the average architecture of residue packing in protein structures[10].

Despite its many successes, the $3 \times 3 \times 3$ cubic lattice model has several drawbacks. In particular, the cubic geometry is a very unlikely one for a polymer. Moreover, with such a geometry, one of the most spectacular property of proteins, namely, their ability to fold into a given structure within a limited amount of time, has to be studied during the course of Monte Carlo-Metropolis processes[9, 11], that is, with a set of elementary displacements chosen *a priori*. Noteworthy, with standard sets of displacements, it is difficult for a polymer to jump from a compact conformation to another, since such a collective motion, during which several monomers are displaced, has to be performed with local moves only, one or two monomers at a time. To overcome this latter difficulty, and to allow for the study of the folding process during the course of Molecular Dynamics simulations, it is necessary to consider off-lattice models. However, with such models, one of the most appealing advantage of the $3 \times 3 \times 3$ cubic lattice model is usually lost: the ability to know, after some simple enumeration, which is the ground state and the energy gap of any given sequence. The off-lattice model shortly introduced in this section has been designed so as to retain most of the advantages of the $3 \times 3 \times 3$ cubic lattice model. In this model, the polymer is represented as a chain of N Lennard-Jones spheres of radius $\frac{R}{2}$, linked by bonds of length R, and the sequence of the polymer is specified by ϵ_{ij}, the well depth of the interaction energy between two monomers. For instance, one can consider the following case: $\epsilon_{ij} = E_{ij}$, where E_{ij} is the same as in lattice models. Thus, the energy function is as follows:

$$H = \sum_{i<j} -(E_{ij} + E_c)\left(\frac{R}{r_{ij}}^{12} - 2\frac{R}{r_{ij}}^6\right)$$

where r_{ij} is the distance between monomers i and j, and where E_c is a "compaction term", large enough to make sure that the lowest energy conformations can be only one among the most compact. Such a term was also used in order to study the folding process of 27-mers on a cubic lattice[11]. While in this latter case the most compact geometry is the $3 \times 3 \times 3$ cubic one, in the case of N Lennard-Jones spheres, though usually more complicated, it is often a very well defined one[14], like in the $N = 19$ case, shown in Fig. 2.

For such a family of models also, all the hamiltonian paths can be determined, as well as the energy gap of any given sequence, and folding simulations of large gap sequences can be performed. Such simulations are currently under way.

References

1. E. Bornberg-Bauer. How are model protein structures distributed in sequence space? *Biophys. J.*, 73(5):2393–2403, 1997.
2. H. Chan and K. Dill. Compact polymers. *Macromolecules*, 22(12):4559–4573, 1989.
3. M. Chorev and M. Goodman. Recent developments in retro peptides and proteins -an ongoing topochemical exploration. *Trends Biotechnol.*, 13:438–445, 1995.
4. M. Ejtehadi, N. Hamedani, H. Seyed-Allaei, V. Shahrezaei, and M. Yahyanejad. Stability of preferable structures for a hydrophobic-polar model of protein folding. *Phys. Rev. E*, 57(3):3298–3301, 1998.
5. N. Go. The consistency principle in protein-structure and pathways of folding. *Adv. Biophys.*, 18:149–164, 1984.
6. H. Li, R. Helling, C. Tang, and N. Wingreen. Emergence of preferred structures in a simple model of protein folding. *Science*, 273:666–669, 1996.
7. H. Li, C. Tang, and N. Wingreen. Nature of driving force for protein folding: A result from analyzing the statistical potential. *Phys. Rev. letters*, 79(4):765–768, 1997.
8. H. Li, C. Tang, and N. Wingreen. Are protein folds atypical? *Proc. Natl. Acad. Sci. USA*, 95(9):4987–4990, 1998.
9. V. Pande and D. Rokhsar. Folding pathway of a lattice model for proteins. *Proc. Natl. Acad. Sci. USA*, 96(4):1273–1278, 1999.
10. G. Raghunathan and R. Jernigan. Ideal architecture of residue packing and its observation in protein structures. *Protein Sci*, 6(10):2072–2083, 1997.
11. A. Sali, E. Shakhnovich, and M. Karplus. How does a protein fold ? *Nature*, 369:248–251, 1994.
12. M. Shemyakin, Y. Ovchinnikov, and V. Ivanov. Topochemical investigations of peptide systems. *Angew. Chem. Int. Ed. Engl.*, 8:492–499, 1968.
13. G. Trinquier and Y.-H. Sanejouand. New protein-like properties of cubic lattice models. *Phys. Rev. E*, 59(1):942–946, 1999.

14. D. Wales and J. Doye. Global optimization by basin-hopping and the lowest energy structures of lennard-jones clusters containing up to 110 atoms. *J. Phys. Chem. A*, 101:5111–5116, 1997.

List of Participants

1. Adams Dean, Iowa State University (USA),
   dcadams@iastate.edu
2. Adebiyi Ezekiel, Universitaet Tuebingen (DE),
   adebiyi@informatik.uni-tuebingen.de
3. Bernasconi Anna, Ist. Matematica Comput. CNR (IT),
   bernasconi@imc.pi.cnr.it
4. Bertonati Claudia, "La Sapienza " University of Rome (IT),
   claudia@medea.ccr.rm.cnr.it
5. Bertrand Hugues-Olivier, University of Rennes (FR),
   hbertrand@msi-eu.com
6. Bock Mary Ellen, Purdue University (USA),
   mbock@stat.purdue.edu
7. Bockmayr Alexander, University Henri Poincare (FR),
   Alexander.Bockmayr@loria.fr
8. Bortoluzzi Stefania, University of Padova (IT),
   stefibo@bio.unipd.it
9. Bosotti Roberta, Pharmacia Upjohn (IT),
   Roberta.Bosotti@eu.pnu.com
10. Califano Andrea, IBM Yorktown (USA),
    acal@us.ibm.com
11. Chiusano Maria Luisa, Second University of Naples (IT),
    marilu@crisceb.area.na.cnr.it
12. Dill Ken, University of California, San Francisco (USA),
    dill@maxwell.ucsf.edu
13. Fransson Linda, Royal Institute of Technology (Sweden),
    lindaf@biochem.kth.se
14. Fuchs Bernhard, University of Bonn (DE),
    fuchsb@cs.uni-bonn.de
15. Houstis Nicholas, Harvard/MIT (USA),
    houstis@genome.wi.mit.edu

16. Jensen Lars, Juhl Technical Univ. of Denmark (DK),
    ljj@cbs.dtu.dk
17. Kaderali Lars, Cologne University (DE),
    kaderali@zpr.uni-koeln.de
18. Kann Maricel, University of Michigan (USA),
    kann@ncbi.nlm.nih.gov
19. Kannan Natarajan, Indian Institute of Science (INDIA),
    kannan@mbu.iisc.ernet.in
20. Kein Holger, Universität zu Köln (DE),
    holger.klein@uni-koeln.de
21. Ketterle Christophe , Genset S.A. (FR),
    Christophe.Ketterle@genset.fr
22. Khyvon Karine, City University, London (UK),
    kyvon@soi.city.ac.uk
23. Kutluay Esin, Cornell Med School (USA),
    esk2001@mail.med.cornell.edu
24. Galbi David, Galbi Research (USA),
    david_ galbi@yahoo.com
25. Gaither Larry, Univ. of Missouri-Columbia (USA),
    lagc4b@missouri.edu
26. Giordano Suzanne, SUNY Stony Brook (USA),
    scgiorda@cedarcrest.edu
27. Goldstein Steve, Genetics Computer Group (USA),
    steveg@gcg.com
28. Grimm Vera, University of Cologne (DE),
    vera.grimm@uni-koeln.de
29. Guerra Concettina, University of Padova (IT),
    guerra@dei.unipd.it
30. Istrail Sorin, Celera Genomics, Rockville, MD (USA),
    sorin.istrail@celera.com
31. Lancia Giuseppe, University of Padova (IT),
    lancia@dei.unipd.it
32. Lesk Arthur, University of Cambridge Clinical School (GB),
    aml2@mrc-lmb.cam.ac.uk
33. Levitt Michael, Stanford University School of Medicine (USA),
    michael.levitt@stanford.edu
34. Lonardi Stefano, University of Padova (IT),
    stelo@artemide.dei.unipd.it
35. Lundegaard Claus, University of Copenhagen (DK),
    clundegaard@strubix.dk
36. Macchiarulo Antonio, University of di Perugia (IT),
    antonio@asia.chimfarm.unipg.it
37. Martin Oliver, University of Cologne (DE),
    oliver.martin@uni-koeln.de

38. Miller David, William Paterson University (USA),
    damiller@pilot.njin.net
39. Moult John, University of Maryland Biotechnology Institute (USA),
    moult@umbi.umd.edu
40. Nicolotti Nicola, University of Bari (IT),
    nicolotti@farmchim.uniba.it
41. Pavesi Giulio, University of Milano (IT),
    pavesi@disco.unimib.it
42. Pellegrini Marialuisa, University of Bari (IT),
    pellegrini@farmchim.uniba.it
43. Quine John R., Florida State University (USA),
    quine@zeno.math.fsu.edu
44. Rantanen Ville-Veikko, University of Turku (Finland),
    vrantane@btk.utu.fi
45. Rossi Andrea, SISSA/ISAS (IT),
    rossi@cm.sissa.it
46. Ruth Erich, University of Miami (USA),
    eruth@cs.miami.ed
47. Schliep Alexander, University of Cologne (DE),
    schliep@zpr.uni-koeln.de
48. Schoenhuth Alexander, Cologne University (USA),
    schoenhuth@zpr.uni-koeln.de
49. Seno Flavio, University of Padova (IT),
    seno@pd.infn.it
50. Shatsky Maxim, Tel-Aviv University (Israel),
    maxshats@math.tau.ac.il
51. Shin Whanchul, Seoul National University (KR),
    nswcshin@plaza.snu.ac.kr
52. Shlyakhova Marianna, Cornell University (USA),
    mis2024@mail.med.cornell.edu
53. Sjunnesson Fredrik, Lunds University (Sweden),
    fredrik.sjunnesson@thep.lu.se
54. Tiuryn Jerzy, Warsaw University (Poland),
    tiuryn@mimuw.edu.pl
55. Zimmermann. Olav, University of Cologne (DE),
    o.zimmermann@uni-koeln.de
56. Wolfson Haim J. , Tel Aviv University (Israel),
    wolfson@post.tau.ac.il